CHAPMAN & HALL/CRC
Monographs and Surveys in
Pure and Applied Mathematics   **105**

PAINLEVÉ

ANALYSIS

AND

ITS APPLICATIONS

CHAPMAN & HALL/CRC
Monographs and Surveys in
Pure and Applied Mathematics   **105**

PAINLEVÉ

ANALYSIS

AND

ITS APPLICATIONS

A. ROY CHOWDHURY

CHAPMAN & HALL/CRC

Boca Raton   London   New York   Washington, D.C.

Library of Congress Cataloging-in-Publication Data

Catalog record is available from the Library of Congress.

Dedication

This book is dedicated to my students and scholars who gave me the opportunity to learn this subject.

Dedication

Contents

CONTENTS

Preface

Study of nonlinear integrable systems has drawn the attention of researchers all over the world for the last three decades. Tremendous development was observed due to the contribution of workers from various disciplines of science. One of the most important tools for the investigation of such systems is the Painlevé analysis. Although a rigorous treatment is quite difficult, it is not impossible to get knowledge to venture into the world of nonlinear systems. This book tries to expose the salient features of the formalism through some important examples which may encourage the reader for a more serious study.

In this respect, I would like to express my indebtedness to various publishers (Springer-Verlag, Elsevier Science, etc.) for allowing me to use different published articles within this book. This book owes its origin to the inspiration of Professor Alan Jeffrey, without whose encouragement it could never have materialized. The staff of CRC Press and Chapman & Hall did an excellent job in molding the book from a raw manuscript into what it is now. Thanks are especially due to Sara Seltzer, Lynn Kelly, Mary Lince, and Helena Redshaw without whose sincere and untiring efforts nothing would have been done. Lastly, it is a pleasure to thank my wife Kasturi and son Anirban for their support which made the task of writing this book a pleasurable one.

A. Roy Chowdhury
Calcutta, India

Chapter 1

Nonlinearity and integrability

1.1 Soliton and nonlinearity

Eugene Wigner in his essay "The Unreasonable Effectiveness of Mathematics in Natural Sciences" pointed out that the chief role of mathematics in physics consists not in its being an instrument (e.g., computations) but in being the language of physics. [1]

This role of mathematics has been served for about the past few hundred years chiefly by differential equations. The general practice is to formulate the laws of physics in the form of differential equations and then solve the differential equations in different physical situations. Though the process of getting the solutions of differential equations is just the inverse of the process of the formation of those equations, it is for several reasons much more difficult to get the solutions. Thus, getting the solutions of differential equations has become a central problem of theoretical physics.

It was not far back when the nonlinear partial differential equations were something of a closed book. The reason is that such equations are very difficult to study. Linear differential equations have the advantage that the principle of (linear) superposition holds in their cases, i.e., by adding two or more solutions. One can always get a new solution, and the general solution can be expressed as a linear combination of the particular solutions. The nonlinear differential equations do not obey the principle of linear superposition. This is a severe loss on the part of nonlinear differential equations and obtaining general exact solutions for the nonlinear differential equations becomes more complicated. However, there are classes of nonlinear (and even linear) equations that possess nonlinear superposition principles. Of course, there is no universal nonlinear superposition.[2,3]

As a result of these difficulties, natural scientists preferred for a long time to use assumptions such as "fluid is inviscid," "given a perfect insulator," "for constant thermal conductivity," or "in a homogeneous and isotropic medium the equations become..." [4] Such assumptions enabled them to keep themselves within the safe zone of linear differential equations. And there was not much urge to reveal the mysteries of nonlinear differential equations.

The boost for the study of nonlinear partial differential equations (nPDEs) started with the work of Zabusky and Kruskal[5] in 1965. Their work was stimulated by a physical problem and is also a classic example of how computational results may lead to the development of new mathematics, just as observational and experimental results have done since the time of Archimedes.[6]

Examining the Fermi-Pasta-Ulm[7] model of phonons in an anharmonic lattice, Zabusky and Kruskal[5,6] were led to the work on the Korteweg–de Vries (KdV) equation. They considered the following initial value problem in a periodic domain:

$$u_t + uu_x + u_{xxx} = 0 \qquad\qquad\qquad (1.1.1a)$$

where

$$u(2,t) = u(0,t) , \qquad u_x(2,t) = u_x(0,t) ,$$
$$u_{xx}(2,t) = u_{xx}(0,t) \qquad\qquad\qquad (1.1.1b)$$

for all t,

and

$$u(x,0) = \cos \pi x \quad \text{for} \quad 0 < x < 2. \qquad\qquad\qquad (1.1.1c)$$

The periodic boundary conditions suit numerical integration of the system. Putting $= 0.022$, Zabusky and Kruskal computed u for $t > 0$. They found that the solution breaks up into a train of eight solitary waves, each like a sech-squared solution, and that these waves move through one another as the faster ones catch up to the slower ones, and finally the initial state (or something very close to it) recurs. The only effect on the interaction is a phase shift, i.e., the center of each wave is at a different position than where it would have been if it had been traveling alone. Because of the analogy with particles, Zabusky and Kruskal referred to these special waves as "solitons."[8]

The history of such studies may be considered to have started in August 1834, when Scott Russell, an experimentalist, observed a large solitary elevation – a rounded, smooth and well-defined heap of water – which continued its course along the channel apparently without change of form or diminution of speed. After 1834 Russell pursued this subject experimentally and apparently tried to interest mathematicians. Indeed, his observations created somewhat of a controversy over whether the equations of motion of fluid mechanics would always produce breaking solutions (see, for example, Airy[9]). The controversy surrounding the idea of solitary wave as Russell conceived it ceased in 1895 when Korteweg and de Vries[10] derived an equation governing moderately small, shallow water waves and thereby found examples of steady traveling (nonbreaking) waves. From 1876 to 1895, several papers contributed to its discussions (see, for example,

McCown[11,12]). However, despite the early work of Korteweg and de Vries,[10] apparently no new application was discovered until 1960.[8] It was then that Gardner and Morikawa,[13] derived the Korteweg-de Vries equation from a study of the collision-free hydromagnetic waves. Finally, in 1965 a new era started with the contribution of Zabusky and Kruskal.[5]

To understand how solitons may persist, we turn to the survey work of Drazin.[6] We take the KdV equation in the form

$$u_t + (1+u)u_x + u_{xxx} = 0 \qquad\qquad (1.2)$$

and seek the properties of small amplitude waves. After linearization, the equation reduces to

$$u_t + u_x + u_{xxx} = 0 \qquad\qquad (1.3)$$

approximately.

Using the method of normal modes, with independent components $u \propto e^{i(kx-wt)}$, it follows from the search for a solution of (1.3) in terms of the superposition of Fourier components that

$$w = k - k^3 . \qquad\qquad (1.4)$$

This is the "dispersion relation," which gives the frequency w as a function of the wave number k. From it, we deduce the phase "velocity,"

$$c = w/k = 1 - k^2 , \qquad\qquad (1.5)$$

which gives the velocity of the wave fronts of the sinusoidal mode. We also deduce the "group velocity"

$$c_g = -dw/dk = 1 - 3k^2 , \qquad\qquad (1.6)$$

which gives the velocity of a wave packet, i.e. a group of waves with nearly the same length $2\pi/k$. It may be noted that $c_g < c < 1$, and $c = c_g = 1$ for long waves (i.e., for $k = 0$). Also a short wave has a negative phase velocity c.

Packets of waves of nearly the same length propagate with the group velocity, individual components moving through the packet with their phase velocity. It can, in general, be shown that the energy of a wave disturbance propagates at the group velocity, not the phase velocity. Long-wave components of a general solution travel faster than the short-wave components, and thereby the components disperse. Thus, the linear theory predicts the dispersal of any disturbance other than a purely sinusoidal one. Looking back to the equation, one can see that the dispersion comes from the term in k^3 in the expression for w and thence from the term u_{xxx} in the KdV equation (1.2).

For further reading on group velocity, Lighthill's book[14] may be consulted.

In contrast to dispersion, nonlinearity leads to the concentration of a disturbance. To see this, the term u_{xxx} can be neglected in the KdV equation (1.2) retaining the nonlinear term. Then we have

$$u_t + (1+u)u_x = 0 \ . \tag{1.7}$$

The method of characteristics may be used to show that this equation has the elementary solution

$$u = f\{x - (1+u)t\}$$

for any differentiable function f. This shows that disturbances travel at the characteristic velocity $(1+u)$. Thus, the higher parts of the solution travel faster than the lower. This "catching up" tends to steepen a disturbance until it breaks and discontinuity or shock wave forms.

For further reading on wave breaking, the books by Landau and Lifshitz[15] and Whitham[16] are very helpful.

One can anticipate that for a solitary wave, the dispersive effects on the terms u_{xxx} and the concentrating effects of the term uu_{xx} are just in balance. Such a balance makes these solutions possible.

Interested readers may consult two survey articles by Miura[17] and Scott et al.[18]

It may be noted that the above developments did not end with the KdV equation and its solitary solutions. A keen interest prevailed and is still continuing among scientists and mathematicians all over the world. The interest is in the understanding of nonlinear differential equations. The next section summarizes some of the important approaches that one can use to extract information regarding a given nPDE (representing a nonlinear wave).

At this point, it may be noticed that other than the solution, another great discovery was made with the help of a computer experiment that radically changed the thinking of scientists about the nature of nonlinearity and introduced new theoretical constructs into the field of dynamics. It is the strange attractor. It is related to the subject of self-generated chaotic behavior in simple dynamical systems (represented by nonlinear ordinary differential equations). The subject is widely recognized as one of the most interesting and intensively studied areas of mathematical physics.[19-26] We will not go into the details of this area. However, several new and exciting results on the chaotic properties of dynamical systems have been obtained by studying Painleve property. Those will be discussed in Section 1.2.

1.2 Some of the important approaches for getting information regarding a given nPDE

1. One can seek steady-state solutions,[6] i.e., waves of permanent form. A soliton is such a solution, although such a solution is, in general, not a soliton.

2. One can see similar solutions.[6,27] They may be found through dimensional analysis or with the group of transformations from one dynamically similar solution to another.

3. One can seek this group and other groups of transformations[6,27,28] under which the nonlinear system is invariant. They are likely to underlie the character of all methods of a solution of a system.

4. One can see as many conservation laws of[24] the given system as possible. An infinity of conservation laws seems to be associated with the existence of soliton interactions.

5. One can seek a Hamiltonian representation[30] of the given nonlinear system. Of course, a partial differential system would have an infinite dimensional Hamiltonian representation or none at all.

6. One can seek the Lax representation[31] of the given nonlinear partial differential equation. In this representation, the given nPDE comes out as an integrability condition of the linear equations.

    This identification is more an art than a science because it depends upon the use of trial and error rather than an algorithm.

    However, there are at least two systematic procedures that work in many cases. The first one is due to Ablowitz *et al.*[32] They leave some arbitrary coefficients in the linear Lax equation and then determine to arbitrary coefficients which are compatible with the integrability condition of the linear equations. And naturally the corresponding nonlinear partial differential equations are also found out. The work of Ablowitz *et al.* is based on the pioneering work of Zakharov and Shabat.[32c] Another procedure is due to Wahlquist and Estabrook.[33] In essence, they try to force the nonlinear equation of interest to be an integrability condition of two linear equations containing the unknown variable and its x derivatives as coefficients. In doing so, they obtain an infinite dimensional algebra or, to put it another way, a set of commutation relations that are not closed. One can close the algebra with the help of symmetry of the original nPDE [33b,33c] or with the help of some *ad hoc* procedure analyzed by Shadwick.[32a]

Once Lax-pair is obtained, one can go for solutions using Inverse Scattering Transform (IST)[32] or using the Riemann-Hilbert problem.[34]

7. One can seek a relevant Backlund transformation of the nPDE.[35] Backlund transformations were shown to be closely associated with the method of inverse scattering [35c,35d] and to be useful in finding multisoliton solutions.

8. One can use the formalism of Ablowitz, Ramani, and Segur[36] to determine whether all ordinary differential equations (ODEs) derivable from the given nonlinear partial differential equations have Painleve property.

9. One can use the formalism of Weiss, Tabor, and Carnevale[37] to determine whether the given nPDE passes the Painleve property in the sense of Weiss *et al.*[37] One can also try to obtain the Lax pair, the Backlund Transform, rational solutions, etc. In recent times, this approach has been attracting much more interest.

10. To obtain exact solutions in a straightforward manner, one can use Hirota's method.[38] However, recent findings indicate that Hirota's method plays a much more central role in the theory than once believed.[18c] It is also becoming more apparent that the methods due to Hirota and Painleve property are closely related.[38b,38c,18c] One can check this in relation with the particular nPDE.

The above list does not provide all possible approaches towards a given nPDE, but it is believed that they are the most prominent. The connections among the findings obtained through different approaches are not very clear even today. For a guideline in this direction, one can see the work of Newell.[18c]

Chapter 2

Basic ideas and methods

2.1 Painleve analysis

An analysis that checks whether a given nonlinear differential equation has Painleve property is generally termed Painleve Analysis. The Painleve Analysis in the sense of Weiss, Tabor, and Carnevale [37] (WTC) for the nonlinear partial differential equations (nPDEs) representing nonlinear waves can be viewed as a natural extension of the Painleve Analysis for nonlinear ordinary differential equations (nODEs) representing dynamical systems. The details are discussed below.

Before elaborating on the theory, consider a very simple nODE whose exact solution can be obtained in terms of elliptic functions. A simple change of this equation gets us to the Painleve equation. The analytic structure and the series form of the solution all can be matched with that of the exact solution. The study of singularity in the complex domain has played a dominant role in the development of the theory of differential equation, be it linear or nonlinear. As for the linear equations, we have the well-known Fuchsian theory, which establishes that once the singularity structure of the functions occurring in differential equation are known, then the solution's analytic structure is also determined. On the other hand, in case of nODEs, the situation is totally different. To understand the problem, it is useful to study an equation closely related to one originally studied by Painleve. This equation is a simpler form of a number of the so called Painleve transcendents. It is written as

$$\frac{d^2 y}{dx^2} = 6y^2 \ .$$

(2.1.1)

We first prove that the exact solution of this equation is written as

$$y = c^2 \left[\frac{-k^2}{1+k^2} + \frac{1}{Sn^2 \{C(x - x_1), k\}} \right] ,$$

(2.1.2)

where C and x_1 are arbitrary constants and k^2 is a root of the equation

$$1 - k^2 + k^4 = 0 \,,$$

Sn being the elliptie function. To proceed, we assume that a solution is

$$y = A + \frac{B}{Sn^2(Cv)} \cdot v = x - x_1 \,.$$

So,

$$\frac{d^2 y}{dx^2} = - \frac{2BC^2}{Sn^6(Cv)} \cdot \left[Sn^3(Cv) \frac{d^2}{dx^2} Sn(Cv) - 3Sn^2(Cv)x \left\{ \frac{d}{dx} \cdot Sn(Cv) \right\}^2 \right] \,. \quad (2.1.3)$$

We now use the identities

$$\left\{ \frac{d}{dx} Snx^2 \right\}^2 = Cn^2x \, dn^2x = \left(1 - Sn^2x\right)\left(1 - k^2 Sn^2x\right)$$

$$\frac{d^2}{dx^2} \cdot Snx = 2k^2 Sn^3x - \left(1 + k^2\right)Snx$$

to reduce equation (2.1.3) to

$$\frac{d^2 y}{dx^2} = \frac{6BC^2}{Sn^4(Cv)} + 2k^2 BC^2 - \frac{4BC^2\left(1 + k^2\right)}{Sn^2(Cv)} \,.$$

Substituting in (2.1.1), we get

$$\frac{6BC^2}{Sn^4(Cv)} + 2k^2 BC^2 - \frac{4BC^2\left(1 + k^2\right)}{Sn^2(Cv)} = 6A^2 + \frac{6B^2}{Sn^4(Cv)} + \frac{12AB}{Sn^2(Cv)} \,.$$

Identifying coefficients, we get

$$6A^2 = 2k^2 BC^2$$
$$6B^2 = 6BC^2$$
$$-6BC^2\left(1 + k^2\right) = 12AB \,,$$

which immediately leads to

$$B = C^2 , \quad A = -\frac{1}{3}C^2\left(1+k^2\right) = \frac{\sqrt{3}}{3}kC^2 ,$$

from which it is easy to deduce that k satisfies

$$1 - k^2 + k^4 = 0 .$$

It is interesting to note that we can also write a solution of the same equation as

$$y = \mathcal{P}(x)$$

where $\mathcal{P}(x)$ is the Weierstrass elliptic function. Another important point is that the solution written above can also be taken in the following form:

$$y = C^2\left[-\frac{k^2}{1+k^2} + k^2 Sn^2(Cv',k)\right]; \quad v' = x - x_0 ,$$

where x_0 is another constant. We now ask the important question, is it possible to analyze the analytic structure of the solution even if we cannot obtain the solution by quadrature? In the case presented above, due to the explicit nature of the solution obtained in terms of known transcendental function, we have the full information of its pole structure and of the number of arbitrary constants involved. In the following, we show that these two important facts about the solution the pole structure and the number of arbitrary constants involved can be obtained even by an infinite series solution. Since equation (2.1.2) shows that y has a pole of second order in $v = x - x_1$, we assume a Laurent series to y as follows:

$$y = \frac{a_{-2}}{v^2} + \frac{a_{-1}}{v} + a_0 + a_1 v + a_2 v^2 + \dots .$$

When we substitute in (1) and equate equal powers of v, we get

$$a_{-2} = 1 , \quad a_{-1} = a_0 = a_1 = a_2 = a_3 = 0$$

$$a_4 = h , \quad a_5 = a_6 \dots = a_9 = 0$$

$$a_{10} = \frac{h^2}{13} a_{11} = a_{12} \dots = a_{15} = 0 \text{ etc.,}$$

with h being an arbitrary constant.

So the series turns out to be

$$y = \frac{1}{v^2} + hv^4 + \frac{h^2}{13}v^{10} + \frac{h^3}{247}v^{16} + \dots \; .$$

We can compare this with the solution (2.1.2). For that, note

$$Snz = z + A_1 z^3 + A_2 z^5 + \dots \; ,$$

whence

$$\frac{z^2}{Sn^2 z} = \frac{1}{C^2 v^2} + C_1 + C_2 C^2 v^2 + C_3 C^4 v^4 + \dots \tag{2.1.4}$$

$$C_1 = 1/3\left(1 + k^2\right)$$

$$C_2 = \frac{1}{15}\left(1 - k^2 + k^4\right)$$

$$C_3 = \frac{1}{189}\left(2 - 3k^2 - 3k^4 + 2k^6\right)$$

$$C_4 = \frac{1}{675}\left(1 - k^2 + k^4\right)^2$$

$$C_5 = \frac{3}{11}C_2 C_3 \text{ etc.}$$

Equation (2.1.4) helps us to identify the Laurent series solution with the exact form (2.1.2), so we can make the important observation that although the equation has no singularities in the finite part of x plane, the solution has a second order pole at $x = x_1$, an important difference from the linear problem. Since the position of the pole x_1 is arbitrary, we say that we have a movable pole. But in the usual practice, one searches for a solution that is analytic in the neighborhood of $x = x_0$ and assumes prescribed values of y and y' say, y_0 and y_0'. That such a solution is also possible follows from the equivalence of the form (2.1.2) with the structure in equation (2.1.4). Since $Sn(x)$ is analytic in the neighborhood of $x = 0$, $Sn(Cv, k)$ will be near $x = x_0$. We can have a Taylor expansion of y as

$$y = y_0 + y_0'(x - x_0) + \frac{y_0''}{21}(x - x_0)^2 + \ldots \, . \tag{2.1.5}$$

Suppose y_0 and y_0' are specified, then y_0'' can be obtained from the equation, y_0''' by differentiating (2.1.5), and so on. We can now obtain two equations connecting the values (v_0, k) with (y_0, y_0'). These are

$$y_0 = \frac{1}{v_0^2} + h v_0^4 + \frac{h^2}{13} v_0^{10} + \frac{h^3}{247} v_0^{16} + \ldots$$

$$y_0' = -\frac{2}{v_0^3} + 4 h v_0^3 + \frac{10}{13} h^2 v_0^9 + \frac{16}{247} h^3 v_0^{15} + \ldots \, .$$

We have a peculiar situation that every point in the x plane can be a polar singularity and yet each point can be a regular point. This is a really paradoxical situation. One can also construct a solution from a very similar equation:

$$\frac{d^2 y}{dx^2} = Ay + By^3$$

in the same manner and obtain

$$y = C \, Sn(\lambda v, k) \cdot v = x - x_0$$

$$k^2 = \frac{\lambda^2 + A}{\lambda^2}, \quad c^2 = -\frac{2(\lambda^2 + A)}{B} \, .$$

We now write down the original equation Painleve wrote down:

$$\frac{d^2 y}{dx^2} = 6y^2 + \lambda x \, . \tag{2.1.6}$$

λ is an arbitrary parameter. Equation (2.1.6) is known as the first Painleve transcendent. For $\lambda = 0$ we get back our equation (2.1.1). We can also apply the same trick here:

$$y = \frac{a_{-2}}{v^2} + \frac{a_{-1}}{v} + a_0 + a_1 v + a_2 v^2 + \ldots \; ; \qquad v = x - x_1 \qquad\qquad (2.1.7)$$

with

$$a_{-2} = 1, \qquad a_{-1} = a_0 = a_1 = 0$$

$$a_2 = \frac{x_1}{10}, \qquad a_3 = -\!/6, \qquad a_4 = h, \quad \text{etc.}$$

One can also have the recussion formula

$$a_n = \frac{6}{n^2 - n - 12} \sum_{k=-1}^{n-3} a_k a_{n-k-2}, \qquad n > 4 .$$

So the series (2.1.7) gives the expansion of the solution in the neighborhood of the singular point ax_1 and can be used effectively in computing values of y near the pole. On the other hand, if one specifies the values y_0, y_0 at a regular point x_0, one can proceed using Taylor expansion and construct two equations:

$$y_0 = 1/v_0^2 - \lambda x_0 /10 \cdot v_0^2 - \lambda /15 \cdot v_0^3 + \ldots$$

$$y_0 = -2/v_0^3 - \lambda x_0 /5 \cdot v_0 - \lambda /5 \cdot v_0^2 + \ldots$$

for the determination of h and v_0 in terms of y_0 and y_0'.

Since the solution of the equation

$$\frac{d^2 y}{dx^2} = 6y^2$$

can be written as

$$y = a + \frac{b}{Sn^2(Cv)} ; \qquad v = x - x_1 ,$$

it appears that one can define a new transcendental function $S(z)$ by

$$S^2(Cv) = \frac{B}{y(c) - A} ,$$

where $y(v)$ is a solution of

$$\frac{d^2y}{dx^2} = 6y^2 + x .$$

We will impose the condition that for $\lambda = 0$, $S(z)$ reduces to $Sn(z)$. So the Painleve equation expands the classes of known transcendental functions. The explicit computation and other properties are given at length in the excellent treatise by H. Davis.

The above analysis of the equation (2.1.1) clearly indicates three important features of a nODE:

1. The solution can have a pole even if the equation does not.

2. The Laurent series solution can be constructed around such a pole.

3. This series will contain the necessary number of arbitrary constants required to specify the solution in the fullest generality.

In the following chapters, these basic properties of various nonlinear ODE and PDE will be discussed. Some variations will also be introduced. To start with, we introduce the α-method originally formulated by Painleve.

Painleve's α-method: [63]
To illustrate this approach, again consider the first Painleve equation:

$$\frac{d^2y}{dt^2} = 6y^2 + a(t) ,$$ (2.1.8)

where $a(t)$ is an analytic function of t. To determine when equation (2.1.8) will have movable critical point, we introduce an arbitrary parameter α by the scaling transformation

$$y = \alpha^{-2}X , \quad t = to + \alpha T .$$

Then we get

$$\frac{d^2X}{dT^2} = 6X^2 + \alpha^4 a(to) + \alpha^5 \frac{da(to)}{dt} + \frac{1}{2}\alpha^6 \frac{d^2a(to)}{dt^2} + 0\left(\alpha^7\right) + ...$$

in which we seek a solution of X in the form of a power series in α,

$$X(T) = X_0(T) + \alpha^4 X^4(T) + \alpha^5 X_5(T) + \alpha^6 X_6(T) + \dots \ . \tag{2.1.9}$$

Note that one can set

$$X(T) = X_0(T) + \alpha X_1(T) + \alpha^2 X_2(T) + \alpha^3 X_3(T) + \alpha^4 X_4(T) + \alpha^5 X_5(T) + \dots$$

instead of (2.1.9). But after the substitution is made in equation (2.1.8), it is observed that X_3, X_4, and X_5 are solved in terms of elliptic functions and are moromorphic, so that they do not affect the determination of which equations of the form (2.1.8) have solutions with no movable critical points. Equating various powers of α we now get

$$\frac{d^2 X_0}{dT^2} = 6X_0^2 \tag{2.1.10}$$

$$\frac{d^2 X_{r+4}}{dT^2} - 12X_0 X_{r+4} = \frac{T^r}{r!} \frac{d^r a}{dt^r}(to) \ . \tag{2.1.11}$$

From our previous discussion, we know that the general solution of (2.1.10) is in terms of Weierstrass elliptic functions. That is

$$X_0(T) = P(T - k, o, h) \ .$$

Now it is known that a general solution of the homogeneous equation

$$\frac{d^2 Y}{dT^2} - 12P(T - k, o, h)Y = 0$$

is

$$Y(T) = \alpha\left(T\frac{dP}{dT} + 2P(T)\right) + \beta\frac{dP}{dT} \ .$$

So by variation of parameter, the solution of equation (2.1.11) is

$$X_{r+4}(T) = U_{r+4}(T)\left(T\frac{dP}{dT} + 2P(T)\right) + V_{r+4}(T)\frac{dP(T)}{dT} \ ,$$

where

$$\frac{dU_{r+4}}{dT} = \frac{T^r}{24r!} \frac{d^r a(to)}{dt^r} \frac{dX_0}{dT}$$

$$\frac{dV_{r+4}}{dT} = \frac{T^r}{24r!} \frac{d^r a(to)}{dt^r} \left(T\frac{dX_0}{dT} + 2X_0 \right). \qquad (2.1.12)$$

Now we know also

$$X_0(T) = P(T-k,o,h) = (T-k)^{-2} + 0\left((T-k)^2\right)...$$

$$\frac{dX_0}{dT} = -2(T-k)^{-3} + 0(T-k) ,$$

whence

$$T\frac{dX_0}{dt} + X_0(T) = -2k(T-k)^{-3} + 0(T-k) .$$

So integrating equations (2.1.12) to obtain U_{r+4}, V_{r+4} for $r = 0$, 1, X_{r+4} is obtained in terms of elliptic functions and so is meromorphic again. For $r = 2$, there must be $\ln(T-k)$ term unless its coefficient is 0. Hence, if the solution is to possess movable critical point, it is necessary that

$$\frac{d^2 a(to)}{dt^2} = 0 .$$

Since to is arbitrary, we should have $d^2 a(to)/dt^2 = 0$ that is $a(t) = a_1 t + a_0$. So we can have three equations with this specific property:

$$\frac{d^2 x}{dt^2} = 6x^2$$

$$\frac{d^2 x}{dt^2} = 6x^2 + 1/2$$

$$\frac{d^2 x}{dt^2} = 6x^2 + t .$$

Painleve test and convergence
Although we will never be seriously concerned about the basic question of convergence of the Painleve series, we mention the simplest situation of the KdV equation for the sake of

completeness of discussion. This particular case was studied by Joshi and Peterson.[62] Incidentally, it should be mentioned that the same question in all six Painleve classes was studied by Joshi and Kruskal. The derivation of the Painleve expansion is an interesting outcome of such an analysis. Let us summarize the results of reference (39) in the following theorem:

Theorem: Let S be holomorphic surface in C^2 given by $\{t = 6\sigma(x)\}$, then locally there exists a solution of the KdV equation

$$u_x + 6uu_t = u_{ttt} \; ,$$

which has the form

$$u(t, x) = \frac{2}{\{t - 6\sigma(x)\}^2} + h(t, x)$$

near S, where $h(t, x)$ is holomorphic. Moreover,

$$\lim_{t \to 6\sigma(x)} \left\{ u_{tt}(t, x) - 3(u(t, x) - \sigma'(x))^2 \right\}$$

and

$$\lim_{t \to 6(x)} \frac{\partial^6}{\partial t^6} \left\{ (t - 6\sigma'(x))^2 u(t, x) \right\}$$

are holomorphic function of x, which can be given arbitrarily.

In the following we will give an outline of the proof following reference (39). The present line of reasoning was influenced by the iteration proof of the Cauchy-Kowalevski theorem given by Nirenberg. It is based on Newton's iteration procedure.

To start with, assume the origin lies on S and straighten it locally via a transformation to the x-plane $(t = 0)$. The KdV equation then becomes

$$u_{tt} = 6(u - \sigma'(x))u_t + u_x \; . \tag{2.1.13}$$

Assume f and g are given holomorphic functions of one variable. Integrate (2.1.13) with respect to t to obtain

$$u_t = 3(u - \sigma'(x))^2 + \int_{t_1}^t dt \, u_x + f(x)$$

t_1 is a point in a neighborhood of S. Successive integration yields,

$$\frac{1}{2} u_t^2 = (u - \sigma'(x))^3 + \int_{t_1}^t d\tau \, u_t \left(\int_{t_1}^\tau d\sigma \, u_x + f(x) \right) + g(x) \qquad (2.1.14)$$

and

$$-\sqrt{2}(u - \sigma'(x))^{-1/2} = \int_{t_1}^t d\tau \left\{ 1 + (u - \sigma'(x))^{-3} \left(\int_{t_1}^t d\sigma u_t \left(\int_{t_1}^\sigma d\rho u_x + f(x) \right) + g(x) \right) \right\}^{-1/2} .$$

$$(2.1.15)$$

Formal iteration of this last equation leads to

$$u = 2/t^2 + \sigma'(x) - (f/10) \, t^2 - \frac{\sigma''(x)}{6} \, t^3 - g(x)/28 \, t^4 + \dots ,$$

which is the same as that obtained by the (WTC) approach. Let us now introduce

$$u = \begin{pmatrix} u_1 \\ u_2 \end{pmatrix} \quad \text{where ;} \quad u_1^2 = \frac{1}{u - \sigma'(x)} \; ; \quad u_2 = \frac{\partial u_1}{\partial t} ,$$

so that (2.1.14) and (2.1.15) can be combined as $\dfrac{\partial U}{\partial t} = FU(t)$, where

$$FU(t) = \begin{pmatrix} U_2 \\ 3\dfrac{u_2^2}{u_1} - \dfrac{3}{2} u_1^{-1} - \dfrac{1}{2} u_1^3 (\bar{f} + \eta) \end{pmatrix} \qquad (2.1.16)$$

with

$$\bar{f} = t\sigma''(x) + f ; \quad \eta = \int_0^t d\sigma \, \partial_x \left(u_1^{-2}(\sigma) \right) .$$

The role of the derivative in Newtonian iteration is played by the linear operator

$$F_u' \omega(t) = \begin{pmatrix} \omega_2 \\ 6 u_2 \omega_2 / u_1 \end{pmatrix} .$$

To solve the KdV, we must integrate (2.1.16), so we set

$$F = \begin{pmatrix} 0 \\ 1/\sqrt{2} \end{pmatrix} + \int_0^t FU(\tau)d\tau ; \quad F_u'\omega(t) = \int_0^t F_u'\omega(\tau)d\tau .$$

Then the desired solution is a fixed point of the operator G in an appropriate function space. The reasoning and the detailed construction of the function space is outside the scope of the present discussion. The interested reader is referred to the original literature for it. The important aspect of the iteration procedure is the derivation of WTC expansion, so one can be assured that there exits a rigorous background of the expansions we use in Painleve analysis.

Difficulties of Painleve analysis

Though Painleve analysis tries to segregate the class of completely integrable systems from the enumerable types of nonlinear systems that we encounter, there are some difficulties associated with the method. It will also make one cautious before drawing any conclusion after applying the test.

2.2 Painleve property and ODEs

The singularities of an ODE can be classified as (1) fixed and (2) movable.[41] While the location of the former is fixed by the nature of the coefficients of the ODE, the latter is a function of the integration constant or initial condition.

Consider the linear first order ODE

$$\frac{dw}{dz} + z^{-2}w = 0 .$$ (2.2.1)

It has the solution, $w = c\exp(1/z)$, c arbitrary const. So, $z = 0$ is the fixed (essential) singular point. Further, the location of this singularity is fixed by the nature of the coefficient of the given ODE.

For a more general case, consider a linear second order differential equation:

$$\frac{d^2w}{dx^2} + p(z) - \frac{dw}{dz} + q(z)w = 0 .$$ (2.2.2)

Its general solution is

$$w(z) = Aw_1(z) + Bw_2(z) \ ,$$

where A and B are two arbitrary constants, depending on the initial data, and $w_1(z)$ and $w_2(z)$ are two independent solutions. The location of the complex singularities of w does not depend on A and B but on the singularities of p and q. They are fixed and independent of the initial values of the problem.

In general, for an nth order linear ODE

$$\frac{d^n w}{dz^n} + P_1(z)\frac{d^{n-1}w}{dz^{n-1}} + \ldots + P_n(z)w = 0 \ ,$$

where $P_i(z)$, $i = 1, 2, \ldots, n$ are all analytic at $z = z_1$ admits n linearly independent solutions in the neighborhood of z, so that the general solution may be written as

$$w(z) = \sum_{i=1}^{n} c_i w_i(z) \ ,$$

where c_is are integration constants. Here, the singularities of the solution must be located at the singularities of the coefficients $P_i(z)$, which are all fixed and *a priori* known and do not depend on the constants of integration c_i, $i = 1, 2, \ldots, n$.

However, if the equation is nonlinear, a different kind of singularity can appear. For example, consider

$$\frac{dw}{dz} + w^2 = 0 \ , \tag{2.2.3}$$

which has the solution

$$w = (z - z_0)^{-1} \ ,$$

where z_0 is a constant of integration.

Thus, at $z = z_0$, w has a singularity, a pole of order one. It is movable because its location depends on z_0.

Similarly,

$$\frac{dw}{dz} + w^3 = 0 , \qquad w = \frac{1}{2}(z-c_1)^{-1/2} \tag{2.2.4a,b}$$

$$-\frac{dw}{dz} + w\log^2 w = 0 , \qquad w = \exp\frac{1}{z-c_1} \tag{2.2.5a,b}$$

$$-\frac{dw}{dz} + \exp(w) = 0 , \qquad w = \log\frac{1}{z-c_1} , \tag{2.2.6a,b}$$

which admit movable algebraic branch point, essential singularity, and logarithmic singularity, respectively. In fact, Fuchs[39a] had shown that the only first order equation that admits no movable critical points (a branch point or essential singularity in the solution of the ODE) is the generalized Riccati equation:

$$\frac{dw}{dz} = P_0(z) + P_1(z)w + P_2(z)w^2 . \tag{2.2.7}$$

The above idea was originally applied by Kovalevskaya (1986)[40] to the equations of motion of a rotating rigid body about a fixed point. By demanding that the solutions of the equations of motion have only movable poles in the complex t-plane, she discovered one more set of parameter values (other than those found by Euler and Lagrange earlier) for which the equations could be integrated and solved exactly in terms of the known functions. It is interesting to note that for all the three choices of parameter values, the equations have only movable poles in the complex t-plane, and no solutions are known when the equations do not have the property, i.e., they have only movable poles in the complex t-plane.

Following the work of Fuchs for first order ODEs, Painleve, Gambier, Garnier, and others[41b,41c] classified 50 nonlinear second order ODEs that do not exhibit critical singularities. Out of these equations, 44 were solved exactly in terms of elementary functions, including elliptic functions. For the remaining six, classes of solutions as well as many properties of their general solution have since been discovered.[41b,41c,43]

Now we are in a position to give precise definition for Painleve property.[36b,41]

1. A critical point is a branch point or an essential singularity in the solution of the ODE.

2.   A critical point is movable if its location in the complex plane depends on the constants of integration of the ODE.

3.   A family of solutions of the ODE without movable critical points have P-property. Here P stands for Painleve.

4.   The ODE is P-type if all its solutions have this property.

Ablowitz, Ramani, and Segur[74] have developed a suitable algorithm to determine whether an ODE (or a system of ODEs) satisfies the necessary conditions to be P-type. In many cases, this algorithm seems to be simpler than the α-method of Painleve and his co-workers,[63] which also determines whether an ODE satisfies the necessary conditions to be P-type. It is rather similar to the method of Kovalevskaya.

Algorithm for a single nth order ODE
For the nth order ODE,

$$\frac{d^n w}{dz^n} = F\left(z, w, \frac{dw}{dz}, \ldots, \frac{d^{n-1} w}{dz^{n-1}}\right) ,$$ (2.2.8)

where F is analytic in z and rational in its other arguments. There are basically three steps to the algorithm.

Step 1:   Find the dominant behavior
Look for a solution of (2.8) in the form

$$w \sim a_0 (z - z_0)^p ,$$ (2.2.9)

where $\text{Re}(p) < 0$ and z_0 arbitrary. Substituting (2.9) in (2.8) shows that for certain values of p, two or more terms may balance (depending on a_0), and the rest can be ignored as $z \rightarrow z_0$. For each such choice of p, the terms that can balance are called the leading terms f (or sometimes, dominant terms). Requiring that the leading terms do balance (usually) determines s_0. If any of the possible p's are not an integer, the equation is not of the P-type. Otherwise, one has to go on with the second step.

Step 2: Find the resonances

If all possible p's are integers, then for each p, (2.2.9) may represent the first term in the Laurent series, valid in a delected neighborhood of a movable pole. In this case, a solution of (2.28) is

$$w(z) = (z - z_0)^p \sum_{j=0}^{w} a_j (z - z_0)^j , \qquad 0 < |z - z_0| < R .$$
(2.2.10)

Here z_0 is an arbitrary constant. If $(n-1)$ of the coefficients $\{a_j\}$ are also arbitrary, these are the n constants of integration of the ODE, and (2.2.10) is the general solution in the deleted neighborhood. The powers of $(z - z_0)$ at which these arbitrary constants enter are called resonances.

For each (p,a) from Step 1, construct a simplified equation that retains only the leading terms of the original equation. Substitute

$$w = a_0 (z - z_0)^p + \beta (z - z_0)^{p+r}$$
(2.2.11)

into the simplified equation. To leading order in β, this equation reduces to

$$Q(r)\beta(z - z_0)^q = 0 , \qquad q > p + r - n .$$
(2.2.12)

If the highest derivative of the original equation is a leading term, $q = p + r - n$, $Q(r)$ is a polynomial of order n. If not, $q > p + r - m$, and the order of polynomial $Q(r)$ equals the order of the highest derivative among the leading terms ($< n$).

The roots of $Q(r) = 0$ determine the resonances

1. One root is always (-1). It corresponds to the arbitrary constant z_0.

2. If a_0 is arbitrary in Step 1, another root is (0).

3. Ignore roots with $\mathrm{Re}(r) > 0$ because they violate the hypothesis that $(z - z_0)^p$ is the dominant term in the expansion near z_0.

4. Any root with $\mathrm{Re}(r) > 0$, but r not a real integer, indicates a (movable) branch point at $z = z_0$. There is no need to continue the algorithm, but it remains to prove that the equation actually has such a branch point.

5. If for every possible (p,a) from Step 1 all the roots of $Q(r)-1$ and possibly (0) are positive real integers, then there are no algebraic branch points. Proceed to Step 3 to check for logarithmic branch points.

6. To represent the general solution of the nth order ODE in the neighborhood of a movable pole, $Q(r)$ must have $(n-1)$ non-negative distinct roots, all real integers. If for every (p,a) from Step 1, $Q(r)$ has fewer than $(n-1)$ such roots, then none of the local solutions is general. This suggests (2.2.9) misses an essential part of the solution.

Step 3:   Find the constants of integration

For a given (p,a) from Step 1, let $r_1 < r_2 < ... < r_s$ denote the positive integer roots of $Q(r)$, $(s < n-1)$. Substitute

$$w = a_0(z-z_0)^P + \sum_{j=1}^{r} a_j(z-z_0)^{p+j} \tag{2.2.13}$$

into the full equation (2.28). The coefficient of $(z-z_0)^{p+j-m}$, which must vanish identically, is

$$Q(j)a_j - R_j(z_0,a_0,a_1...a_{j-1}) = 0 . \tag{2.2.14}$$

1. For j (r_1), (2.2.14) determines a_j.

2. For $j = r_1$, (2.2.14) becomes

$$0 \cdot a_{r_1} - R_{r_1}(z_0,a_0,a_1...a_{r-1}) = 0 .$$

    If
$$R_{r_1}(z_0,a_0,a_1...a_{r-1}) = 0 , \tag{2.2.15}$$

then (2.14) cannot be satisfied. There is no solution of the form (2.13), and we must introduce logarithmic terms into the expansion. Replace (2.2.13) with

$$w = a_0(z-z_0)^P + \sum_{j=1}^{r-1} a_j(z-z_0)^{p+j} + [a_{r_1} + b_{r_1} \ln(z-z_0)](z-z_0)^{p+r_1} + ... \tag{2.2.16}$$

Now the coefficient of $\left\{ (z-z_0)^{p+r_1-n} \ln(z-z_0) \right\}$ is

$$Q(r_1)b_{r_1} = 0 \ ,$$

but b_{r_1} is determined by demanding that the coefficient of $(z-z_0)^{p+r_1-n}$ vanish, a_{r_1} is arbitrary. Continuing the expansion (2.16) to higher orders introduces more and more logarithmic terms. Thus (2.2.15) signals a (movable) logarithmic branch point. If such a series can be proved to be asymptotic as $z \to z_0$, the equation is not P-type.

3. If it happens that (2.2.15) is false (i.e., $R_{r_1} = 0$), then a_{r_1} is an arbitrary constant of integration. Proceed to the next coefficient.

4. Any resonance that is a multiple root of $Q(r)$ represents a (movable) logarithmic branch point with an arbitrary coefficient. If the assumed representation is asymptotic as z z_0, the equation is not p-type.

5. At each nonresonant power, (2.2.14) determines a_j. At each resonance, either

$$R_{r_1} = 0 \ ,$$

logarithmic terms must be introduced in (2.2.13), and the equation is not of P-type, or

$$R_{r_1} = 0$$

and a_{r_1} is an arbitrary constant of integration.

6. If no logarithms are introduced at any of the resonances, one could in principle compute all the terms in the series. However, because the recursion relations are nonlinear, it is usually not feasible to determine the region of convergence of the series as one does in a linear problem. An alternative is to prove directly from the ODE that each arbitrary constant (a_j) is the coefficient of an analytic function (41.b).

7. If no logarithms are introduced at any of the resonances for all possible (p,a) from Step 1, then the equation has met the necessary conditions to be P-type (under the assumption that $p \neq 0$).

Algorithm for a coupled system

The basic steps are not essentially different from what we have already discussed.

Step 1: Dominant behavior of the system

Substitute

$$w_j \sim a_{0j}(z-z_0)^{pj} , \quad j=1,\ldots,n \tag{2.2.17}$$

into the equations of the system and determine $\{p_j\}^n = j = 1$, for which there is a balance of leading terms, and what the leading terms are. Ordinarily, the a_{0j} are not entirely determined, but must satisfy k relations, with k n. The algorithm stops unless the only possible p_j's are integers.

Step 2: Resonances of the system

If all possible p's are integers, then for each p the expansions (2.2.17) may represent the first term in the Laurent series, valid in a deleted neighborhood of a movable pole. The solutions of the system of equations are represented by

$$w_j(r) = (z-z_0)^{pj} \sum_{k=0}^{\infty} a_{kj}(z-z_0)^k . \tag{2.2.18}$$

For each p, construct a simplified equation from the equations of the system that retain only the leading terms. Substitute into this simplified equations

$$w_j = s_{0j}(z-z_0)^{pj} + \beta_j(z-z_0)^{p_j+r} , \quad j=1,\ldots,n \tag{2.2.19}$$

with the same r for every w_j. To leading order in β, this becomes

$$Q(r)p = Q , \tag{2.2.20}$$

where Q is an $n \times n$ matrix, whose elements depend on r. The resonances are the non-negative roots of

$$\det[Q(r)] = 0 \tag{2.2.21}$$

a polynomial of order $< n$. One root is always (-1), which corresponds to arbitrary z_0. Zero may also be a root with multiplicity depending on how many a_j's were determined in Step 1. The algorithm stops unless all the resonances are integers.

Here it may be noted that multiple root at any resonance does not necessarily indicate a logarithmic branch point.

Step 3: The constants of integration
Substitute in the equations of the system

$$w_j = a_{0j}(z - z_0)^{p_j} + \sum_{k=1}^{r_s} a_{kj}(z - z_0)^{p_j + k} \ ,$$

where r_s is the largest resonance. The coefficient of each power of $(z - z_0)$, which must vanish, has the form of a matrix generalization of (2.2.14).

Algorithm for equations (or systems) where complex conjugate of an unknown function appears explicitly [36b]

1. Write the ODE for w, and the ODE for v.

2. Treat $v = \overline{w}$ as a new variable, then apply the algorithm to this system of two ODEs, without assuming any relation between v and w. Because the equation for v is the formal complex conjugate of that for w, if the initial conditions at a point of the real axis are formally complex conjugate, then the solutions will satisfy

$$v(z) = [\overline{w(z)}] \ .$$

The behavior of v at a singular point z_0 is related to the behavior of w at $\overline{z_0}$, but not at z_0 unless z_0 is real. If among the possible leading behaviors there are some where v and w differ, it only means that if complex conjugate initial data is given, there can be no such singularities on the real axis.

2.3 Weak Painleve property and ODEs
Ramani, Dorizzi, and Grammaticos [42] have suggested that the existing ARS algorithm can be generalized by introducing the so-called weak P-property. In the case of weak P-property, certain algebraic branch points are allowed. Here, p' in (2.10) can be a

negative rational number with an integer denominator $d > 0$ $(d \neq 1)$. And for that case, the required expansion is given by

$$w(z) = (z - z_0)^{-q/d} \sum_{j=0}^{\infty} a_j (z - z_0)^{j/d} \ , \qquad (2.2.22)$$

where q and d are positive integers and $d \neq 1$. With this expansion in hand, one has to go through the same steps as in the case of (2.10) or (2.18). $j = -d$ seems always to be a resonance and corresponds to the arbitrary constant z_0.

If one observes that the expansion (2.2.22) admits the same number of arbitrary constants as the number of resonances, then one can say that the expansion (2.2.22) represents the general solution of the equation concerned and has weak P-property. A coupled system having any of p_j rational one has to go through the same procedure as in the case of (2.2.18) with expansions such as (2.2.22) for those w_j having rational (noninteger) values of p_j. As per the idea of Dorizzi $et\ al.$ a system is said to have weak Painleve property whenever the solution in the neighborhood of a singularity $z = z_0$ can be expressed as an expansion in powers of $(z - z_0)^{1/d}$, where d is an integer. In their study of several classes of potentials, they considered homogeneous potentials of degree $(p + 2)$. It is obvious that the leading singularity will be

$$x, y \ \alpha (t - t_0)^{-2/p}$$

where (x, y) are the coordinates of the two dimensional dynamical system. Moreover, there is always a resonance at $2 + 4/p$, which except at $p = 1$, 2, or 4 is not an integer. In general the solution will be an expansion in power in $(t - t_0)^{1/p}$, showing that a possible choice for d is p. By redefining the dependent variables, one may again have the usual Painleve properties. In those cases, the procedure detailed above can always be repeated for such systems, and one can check the existence of the sufficient number of arbitrary expansion coefficients. A straightforward extension is also possible for the partial differential equations. Several instances of such situations will be revealed in the following discussions.

Chapter 3
Conformal invariance

3.1 Introduction

Previous discussions spelled out the basic structure underlying the Painleve analysis. In 1989, Conte[58,60] introduced an elegant imodification that exhibited the invariance of the whole procedure under a conformal transformation. This is due to the important fact that the classification of the various sorts of singularities (pole, zero, branch point, etc.) is obviously invariant under the six-parameter Mobius group, defined as

$$z \rightarrow \frac{az+b}{cz+d} \ , \quad (a,b,c,d) \ \varepsilon \ C \ , \quad ad-ch=1 \ .$$

We have already seen that the Painleve property of ordinary differential equations (ODEs) is basically dependent on such a classification, so it seems that some kind of simplification may result if one invokes such invariance. The extension to partial differential equations (PDEs), which is an extension of the same procedure, should therefore also be invariant under the same transformation. Of course, now the transformation is to operate on $\phi(x,t)$, which defines the singularity manifold $\phi = 0$.

We now recapitulate the basic elements needed to set up the Painleve analysis:

$$\phi(x_1 ... x_n) = 0 \ ,$$

the singular manifold.

$p = \{p_m\}$, one of the possible sets of leading order exponents of the dependent variable u.

$q = \{q_1\}$, the associated sequence of leading exponents of the system of equation.

$(p, u_0) = \{p_m, u_0^m\}$, one of the allowed branches.

$(u_j) = \{u_j^m\}$, the coefficients of the Painleve expansion of u when it is a simple series in ϕ, defined as

$$u^{(m)} = \phi^{Pm} \sum_{j=0}^{\infty} u_j^{(m)} \phi^j \ .$$ (3.1.1)

The following system of equations will be collectively denoted as $F^{(1)} = 0$, and it depends on several variables, such as x_i, u_j, Du_k, etc., which we write as

$$F^{(1)}\left(x_i, u_j^{(m)} \ Du_k\right) = 0$$ (3.1.2)

or in vector notation as

$$\overset{\vee}{F}(X, u, Du) = 0 \ .$$

Let us denote the coefficients of expansion of F as $\left\{F_j^1\right\}$ occurring in

$$F^1 = \phi^{q_1} \sum_{j=0}^{\infty} F_j^1 \phi^j \ .$$ (3.1.3)

It is quite easy to be convinced that such a simple series can exist only if certain conditions $Q_r^{(1)} = 0$ are satisfied, where r runs over the so-called incompatible resonances. On the other hand, if these conditions are not satisfied, then the logarithmic terms must be introduced. The above series is then converted into a double (or in general, a multiple) series:

$$u^{(m)} = \phi^{Pm} \sum_{j=0}^{\infty} \sum_{k=0}^{\infty} u_{jk}^{(m)} \phi^j (\log \phi)^k \ .$$ (3.1.4a)

The main theme of this discussion is that the whole formalism of the Painleve analysis is invariant under the above transformation.

The basic idea was to dissociate the two roles played by the same function ϕ, definition of the singular manifold and the expansion variable. We now use X as the expansion variable and let ϕ be the singular manifold. The function X is assumed to be dependent on ϕ and $D\phi$. The specific form of the function is to be fixed from the basic requirement that the change $\phi \rightarrow X$ should preserve the type of singularity located at

$\phi = 0$. Such a condition can be satisfied if ϕ and X are connected by a homographic transformation with $b = 0$ and a, c, and d in general complex. Now it is well known that

$$\text{grad } X = X_0(D\phi) + X_1(D\phi)X + X_2(D\phi)X^2 . \tag{3.1.4b}$$

So that each component of this is a Riccati type ODE. We now want the three coefficients X_i ($i = 0,1,2$) to have the maximum invariance. Now let G be the largest subgroup of H whose action on ϕ leaves X_i invariant. Then G contains the two-parameter subgroup of translation $\phi \rightarrow \phi + b$, $b \varepsilon C$. We now want our X to be such that G is equal to H. It is then possible to show that for a PDE of the form $F\left(x_i, u_j, DU_k\right) = 0$ (F is polynomial in u and Du), every result of the Painleve analysis using an expansion variable X that conforms to the above two criterion has a special structure that is "invariant under G, algebraic in x_0, polynomial in X_1, X_2, DX_i." In the following we outline a simple proof of the above assertions following Conte.

1. Remember that in the first step of our Painleve analysis we look for the leading power p of u – selected from the dominant terms of F. If they have different degrees, then we get a rational value of p, and their coefficients lead to a polynomial equation in p and X_0. These p are invariant under G, algebraic in X_0. This step also leads to the determination of the leading powers q of F.

2. In the next step we assume u_0, X_0 constant, X_1, X_2 0, and set

    $$u = u_0 X^p$$

    in the dominant terms of $F = 0$. The result is a polynomial system in u_0 and X_0. Its solutions u_0 are invariant under G algebraic in X_0.

3. Next we assume

    $$u = u_0 X^p + u_r X^{r+p}$$
    $$Du = u_0 D\left(X^p\right) + u_r D\left(X^{r+p}\right) \tag{3.1.5}$$

    in the dominant terms of $F = 0$, keeping only terms of F with exponent $r+q$ in X. In the resulting linear, homogeneous system for u_r (with coefficients polynomial in

u_0 and X_0), one requires the determinant to vanish – determining the resonances which are again invariant under G.

4. We now plug in the full series for u

$$u = \sum_{j=0}^{\infty} u_j X^{j+p} \tag{3.1.6}$$

in

$$F\left(x_i, u_j, Du_k\right) = 0 ,$$

which then becomes an equation of the form

$$F\left(x_i, \left(u_j\right), \left(Du_k\right), X, DX\right) = 0$$

a series in X, DX, with coefficients in u_j and Du_j. Then using (3.1.4a) we express every derivative of X as a polynomial in X with coefficients depending on X_i and DX_i. Now G acts on ϕ, not on x_i, so a derivation with respect to x does not destroy the aforementioned invariance so that every derivative DX can be expressed as a polynomial in X with coefficients invariant under G – they being polynomial in X_i, DX_i.

$$DX = \sum_{k=0}^{\infty} N_K\left(X_i, DX_i\right) X^K \tag{3.1.7}$$

Substituting these expressions in $F^{(1)}$, we get the usual recursion relation of u_j and u_{k-j}. So we can infer that even up to the determination of the expansion coefficients the invariance is intact. One may note that this determination involves only some algebraic manipulations – addition, multiplication, derivation, and division – all of which preserve the invariance under G.

5. In the last stage, if p and q both are negative integers, then we truncate the series by setting

$$u_j = 0 , \quad j = -p+1, \ldots -q ,$$

which yields the Backlund transformation. It now remains to make the process more explicit by the choice of X. Note that

$$X = \left(\frac{\phi_x}{\phi} - \frac{\phi_{xx}}{2\phi_x} \right)^{-1} \tag{3.1.8}$$

satisfies both requirements. Also, one can compute and prove very easily

$$X_t = -C + C_x X - 1/2 \left(C_{xx} + CS \right) X^2 \tag{3.1.9}$$

with $S = \dfrac{\phi_{xx}}{\phi_x} - \dfrac{3}{2} \left(\dfrac{\phi_{xx}}{\phi_x} \right)^2$

$$C = -\phi_t / \phi_x \tag{3.1.10}$$

S and C are two homographic invariants. If instead of t we take x derivative, then

$$X_x = 1 + 1/2 \, SX^2 \tag{3.1.11}$$

S is known as the Schwarzian derivative. Also, due to the consistency

$$\left(\phi_t \right)_{xxx} = \left(\phi_{xxx} \right)_t$$

we have the relation

$$S_t + C_{xxx} + 2C_x S + CS_x = 0 \ . \tag{3.1.12}$$

Lastly, we make some comments about the relation between invariant and noninvariant computation. This is obtained from

$$u = \phi^s \sum_{j=0}^{\infty} a'_j \phi^j = X^s \cdot \sum_{k=0}^{\infty} a_k X^k \ . \tag{3.1.13}$$

Using the expression for X,

$$V_j \; ; \quad \phi_x^{j+s} a_j' = \sum_{k=0}^{\infty} C_{j-k,-k-s} \left(- \frac{\phi_{xx}}{2\phi_x} \right)^{j-k} , \qquad (3.1.14)$$

where

$$(1+x)^{\alpha} = \sum_{n=0}^{\infty} C_{n,\alpha} X^n .$$

Before closing the discussion, we should mention the status of the Kruskal assumption $\phi_x = 1$ in this new set up. Note that $\phi_x = 1$ implies

$$S = 0 \; ; \quad C_x = 0 ,$$

from which we at once infer that

$$\phi = \frac{a[x - f(x_n)] + b}{c[x - f(x_n)] + d} \; ; \quad ad - bc = 1 , \qquad (3.1.15)$$

where f is an arbitrary function of the $N-1$ independent varieties x_n other than x, a, b, c, d being complex constants. So we observe that the Kruskal assumption is equivalent to the invariant Painleve analysis along with the constraint (3.1.15). It may be noted that in this form of Painleve analysis, one never has to use the explicit form of X but need only the properties associated with X, X_x, and X_t.

We next illustrate the above discussion with the help of a simple example. It is actually the Mikhailov equation [45]

$$v_{xt} - 2a^2 e^v + b^2 e^{-2v} = 0 .$$

To convert it into a polynomial form, we set $u = e^{-v}$, so that one obtains

$$F = -u u_{xt} + u_x u_t - 2a^2 u + b^2 u^4 = 0 . \qquad (3.1.16)$$

Leading order exponents p can be obtained by equating any two of three different powers $- 2p - 2$, p, and $4p$ – which yields $(-1, 0, 2)$. We choose to study $p = -1$. The value of q is now -4. Using the properties of X, X_x, X_t, etc.,

$$X_x \sim 1; \quad X_t \sim -c, \quad u \sim u_0 X^{-1}, \quad u_x \sim -u_0 X^{-2}$$

$$u_t \sim C u_0 X^{-2}; \quad u_{xt} \sim -2C u_0 X^{-3}$$

with $u_0^2 + b^{-2} C = 0$.

In the next stage, substitute

$$u = u_0 X^{-1} + u_j X^{j-1} \tag{3.1.17}$$

whence the coefficient of u_j is found to be

$$B = (j+1)(j-2) u_0 C \tag{3.1.18}$$

so that resonances exist at $j = -1$, $j = 2$. We also have

$$B^{-1} = -\frac{b^2}{(j+1)(j-2)} \cdot u_0 / C^2 \tag{3.1.19}$$

$$u_{0x} = 1/2 C^{-1} C_x u_0, \quad u_{0t} = 1/2 C^{-1} C_t u_0$$

The only useful compatibility condition is $Q_2 = 0$. One also finds

$$u_1 = -1/2 C^{-1} C_x u_0$$

$$F_3 = -2a^2 u_0 - 1/2 b^{-2} C (s - s_1)_t = 0 \qquad , \tag{3.1.20}$$

$$F_4 = a^2 C^{-1} C_x u_0 + 1/4 b^{-2} \left[C^2 (s - s_1)(s - s_2) + C_x (s - s_1)_t \right] = 0$$

with the notation

$$S_1 = 1/2 C^{-2} C_x^2 - C^{-1} C_{xx} \tag{3.1.21}$$

$$S_2 = S_1 - C^{-3} C_t C_x + C^{-2} C_{xt}$$

Equation (3.1.20) implies $(S - S_1)(S - S_2) = 0$, and the cross derivative condition yields $S_{1,t}$ for $S = S_1$ and to $S_{2t} - S_{1t}$ for $S = S_2$. So one may conclude that the Backlund transformation has no solution except in the case $a = 0$.

3.2 KdV revisited

The Korteweg-de Vries (KdV) problem has already been discussed in detail, but here we consider the same in light of the invariant expansion. The KdV equation is written as

$$F = u_t + uu_x + u_{xxx} = 0 .$$ (3.2.1)

We set $u(xt) = \phi(xt)^p \sum_{j=0}^{\infty} u_j(xt)\phi(xt)^j$, where $p = -2$. Substituted in equation (3.2.1), it leads to

$$\sum_{j=0}^{\infty} F_j\left(u_0,...,u_j, \phi_x, \phi_t, D\phi_x...\right)\phi^{j-5} = 0 .$$

$F_0 = 0$ leads to $u_0 = -12\phi_x^2$ and

$$F_j = (j+1)(j-4)(j-6)\phi_x^3 u_j + Q_j = 0$$

Q_j depends on $u_0,...,u_{j-1}$, ϕ_x, ϕ_t and their derivatives.

The known resonances appear at $j = 4$ and $j = 6$. The Painleve property is characterized by the fact that p is a negative integer, and all resonances occur at positive integers and are compatible. Next the truncation at the constant level ϕ yields

$$u = \phi^{-2} \sum_{j=0}^{2} u_j \phi^j$$ (3.2.2)

$$u_1 = -12\phi_{xx}$$
$$u = 12(\log \phi)_{xx} + u_2$$ (3.2.3)

Equation (3.2.3) yields a Backlund transformation if

$$j = 2 : \phi_x \phi_t + \phi_x^2 u_2 + 4\phi_x \phi_{xxx} - 3\phi_{xx}^2 = 0$$

$$j = 3 : \phi_{xt} + \phi_{xx} u_2 + \phi_{xxxx} = 0$$

$$j = 5 : u_{2t} + u_{2x} u_2 + u_{2xxx} = 0$$

We now use the invariant functions S, C and the noninvariant rational $K = -\dfrac{\phi_{xx}}{2\phi_x}$, and remember the condition given in equation (3.1.9). If we set

$\phi_x = \psi^{-2}$, then

$$K = (\ln \psi)_x \qquad (3.2.4)$$

$$K^2 = -S/2 - (\ln \psi)_{xx}$$

We now define invariant expressions

$$\tilde{F}_j = \sum_{k=3} (-K)^{j-k} \frac{(5-k)!}{(5-j)!(j-k)!} F_k$$

so that expressions (3.24) are replaced by

$$u_2 = C + 2S + 12(\ln \psi)_{xx} \qquad (3.2.5)$$

$$\tilde{E}_3 = 0 : -12(S - C)_x = 0 ,$$

whence $S - C = \lambda(t)$

$$E_5 = 0 : C_t + CC_x - 6SC_x - 5S_t - 5CS_x - 2SS_x - 4S_{xxx} = 0 . \qquad (3.2.6)$$

The condition (3.26) is equivalent with

$$S = 1/3(u' + \lambda) \qquad (3.2.7)$$

if $\dfrac{\partial \lambda}{\partial t} = 0$, and u a solution of (3.2.1).

Now since $\phi_x = \psi^{-2}$, we get

$$S = -2\psi^{-1}\psi_{xx}$$

so that (3.2.5) and (3.2.6) immediately yields

$$\psi_{xx} + 1/6(u' + \lambda)\psi = 0 .$$
(3.2.8)

In a similar fashion, we can also derive

$$\psi_t = 1/3(2\lambda - u')\psi + 1/6u'_x\psi .$$
(3.2.9)

So we have derived the space and time part of the Lax pair in a very simple manner from the Painleve analysis.

Another important feature of this invariant approach is the introduction of three more invariant functions Y_i of two linearly independent solutions ψ_1, ψ_2 of the yet unknown Lax pair in scalar form

$$\frac{Y_1}{W_x(\psi_1, \psi_2, x)} = \frac{Y_2}{W(\psi_{1x}, \psi_{2x}, x)} = \frac{Y_3}{W(\psi_1, \psi_2, t)} = \frac{1}{W(\psi_1, \psi_2, x)} ,$$

with $W(u, w, y) = u_y \omega - u \omega y$.

These three functions Y_i are invariant under the group $SL(2, C)$ whose projective group is

$$\begin{pmatrix} \psi_1 \\ \psi_2 \end{pmatrix} \rightarrow \begin{pmatrix} c & b \\ c & d \end{pmatrix} \begin{pmatrix} \psi_1 \\ \psi_2 \end{pmatrix} ; \quad ad - bc = 1 .$$

The basic expressions needed for the Lax pair have simple dependence on Y_1, Y_2, Y_3. For a second order Lax pair, every Wronskian is proportional to $W(\psi_1, \psi_2, x)$. For Lax pair of order three the variables Y_i introduced so far are enough to define a psuedopotential of projective type Riccati equation with invariant components. For still higher order Lax operator, one needs to define additional homographic invariant quantities.

3.3 Boussinesq equation

We now pass over to a higher order equation associated with a third order Lax operator. This is the Boussinesq equation. The equation is written as

$$u_{tt} = -\frac{\partial^2}{\partial x^2}\left(\frac{u_{xx}}{3} + u^2\right).$$
(3.3.1)

It is easy to ascertain that the expansion of u will be

$$u = \phi^{-2}\sum_{j=0}^{\infty} u_j\phi^j ,$$
(3.3.2)

which is the leading exponent $p = -2$. From the recursion relation, the resonances are seen to be situated at $j = 4, 5, 6$ and, of course, at $j = -1$, so that u_4, u_5, u_6 are arbitrary. The other coefficients are determined to be

$$u_0 = -2\phi_x^2$$

$$u_1 = 2\phi_{xx}$$

$$\phi_t^2 = \phi_{xx}^2 + \frac{4}{3}\phi_x\phi_{xxx} + 2\phi_x^2 u_2 = 0$$

$$\phi_{tt} = \frac{1}{3}\phi_{xxxx} + 2\phi_{xx}u_2 - 2\phi_x^2 u_3 = 0$$
(3.3.3)

By truncation, the Backlund transformation is determined to be

$$u = u_0\phi^{-2} + u_1\phi^{-1} + u_2$$

$$= 2\frac{\partial^2}{\partial x^2}\ln\phi + u_2 ,$$
(3.3.4)

where

$$2u_2 + \frac{\phi_t^2}{\phi_x^2} - \frac{\phi_{xx}^2}{\phi_x^2} + \frac{4}{3}\frac{\phi_{xxx}}{\phi_x} = 0$$
(3.3.5)

along with

$$\frac{\partial}{\partial t}(\phi_t/\phi_x)+1/3\left(\{\phi,x\}+3/2(\phi_t/\phi_x)^2\right)=0 \ , \tag{3.3.6}$$

where

$$\{\phi,x\} = \frac{\partial}{\partial x}\cdot(\phi_{xx}/\phi_x)-1/2(\phi_{xx}/\phi_x)^2 \ ,$$

the Schwarzian derivative which is invariant under the homographic transformation. If we let

$$v = \frac{\phi_{xx}}{\phi_x} \ , \qquad w = \frac{\phi_t}{\phi_x} \ , \tag{3.3.7}$$

then (v,ω) satisfy the modified Boussinesq equation

$$v_t = \frac{\partial}{\partial x}(\omega_x + v\omega)$$
$$t = -1/3\frac{\partial}{\partial x}\left(v_x - \frac{1}{2}v^2 + \frac{3}{2}\omega^2\right) \cdot \tag{3.3.8}$$

The Miura map is given as

$$2u_2 +\omega^2 +4/3\left(v_x +1/4v^2\right)=0 \ . \tag{3.3.9}$$

Since this transformation maps the system (3.3.8) into the scalar equation (3.3.9), it is convenient to reformulate (3.3.1) as the coupled equation

$$u_t = H_x$$
$$H_t = \frac{\partial}{\partial x}\cdot\left(-u_{xx}/3-u^2\right) \ , \tag{3.3.10}$$

with the Miura map

$$2u+\omega^2 +\frac{4}{3}\left(v_x +\frac{1}{4}v^2\right)=0 \tag{3.3.11}$$

$3H + 2\omega_{xx} - \omega^2 +v_x\omega+3v\omega_x +v^2\omega = 0$ is invariant under

$$\begin{pmatrix} v \\ \omega \end{pmatrix} = A_\pm \begin{pmatrix} \theta \\ z \end{pmatrix} ,$$

(3.3.12)

with

$$A_\pm = \begin{pmatrix} -1/2 & \mp 3/2 \\ \pm 1/2 & -1/2 \end{pmatrix} .$$

(3.3.13)

Note that $A_\pm$ possesses the property

(i) $|A_\pm| = 1$

(ii) $A_\pm^{-1} = A_\mp$

(iii) $A_\pm^3 = I$.

It is interesting to analyze the equation (3.3.6) for singularities. Equation (3.3.6) allows two types of singularities. For one

(a) $\phi = \varepsilon^{-1} \sum_{j=0}^{\infty} \phi_j \varepsilon^j$

(3.3.14)

and for the other

(b) $\phi = \phi_0(t) + \phi_2 \varepsilon^2 + ...$

(3.3.15)

with

$$\phi_{0t} = \pm 2\varepsilon_x \phi_2 .$$

Singularities of the second type occur when $\phi_x = 0$. By direct computation it is not difficult to assert that both forms of singularity are single valued. Also, the invariance of (3.3.6) under

$$\psi = 1/\phi$$

throws the simple pole of ϕ into a simple 0 of ψ :

$$\psi = \varepsilon \sum_{j=0}^{\infty} \psi_j \varepsilon^j ,$$

where ψ is locally analytic near $\varepsilon = 0$. We note that by the Cauchy-Kovalevsky theorem converges in an open neighborhood at the manifold $\varepsilon = 0$. For simplicity, let

$$\varepsilon \to x + \varepsilon(t)$$

and find to leading order

(i) $v = \dfrac{\phi_{xx}}{\phi_x} \simeq -2/\varepsilon$

$$\omega = \frac{\phi_t}{\phi_x} \approx 0(1)$$

$$\begin{pmatrix} v \\ \omega \end{pmatrix} = \begin{pmatrix} -2 \\ 0 \end{pmatrix} \cdot \varepsilon^{-1} \quad \text{for (3.3.14)}$$

and

(ii) $\begin{pmatrix} v \\ \omega \end{pmatrix} \approx \begin{pmatrix} 1 \\ \pm 1 \end{pmatrix} \cdot \varepsilon^{-1} \quad \text{for (3.3.15).}$

We now observe that

$$A_\pm \begin{pmatrix} -2 \\ 0 \end{pmatrix} = \begin{pmatrix} 1 \\ 1 \end{pmatrix}, \quad \begin{pmatrix} 1 \\ -1 \end{pmatrix}$$

$$A_\pm \begin{pmatrix} 1 \\ 1 \end{pmatrix} = \begin{pmatrix} -2 \\ 0 \end{pmatrix}, \quad \begin{pmatrix} 1 \\ -1 \end{pmatrix}$$

$$A_\pm \begin{pmatrix} 1 \\ -1 \end{pmatrix} = \begin{pmatrix} 1 \\ 1 \end{pmatrix}, \quad \begin{pmatrix} -2 \\ 0 \end{pmatrix}.$$

Thus, the singularities are permuted by the symmetry.

The Schwarzian derivative

The Schwarzian derivative,

$$\{\phi, x\} = \frac{\partial}{\partial x}\left(\frac{\phi_{xx}}{\phi_x}\right) - 1/2\left(\frac{\phi_{xx}}{\phi_x}\right)^2 ,$$

is a third order differential expression that is invariant in form under the homographic group acting on the dependent variable, that is

$$\left\{\frac{a\phi + b}{c\phi + d} \; ; x\right\} = \{\phi, x\} .$$

Furthermore,

$$\{\phi, x\} = \{\psi, x\} + \{f, \psi\}\psi_x^2$$

$$\{\phi, x\} = h'^2\{\phi, z\} + \{h, x\}$$

for respectively

(a) $\phi = f(\psi)$, an arbitrary change of dependent variables.

(b) $z = h(x)$, an arbitrary change of independent variables.

In this context, it is interesting to consider the equation associated with the KdV-MKdV equation

$$\phi_t/\phi_x + \sigma\{\phi, x\} = \lambda .$$

(3.3.16)

This equation is homogeneous in ϕ and consists of two terms, one ϕ_t/ϕ_x invariant under arbitrary changes of dependent variable $\phi = f(\psi)$, the other a function of the Schwarzian derivative invariant under the homographic group. As a consequence, equation (3.4.6) is invariant under the same group. Following the usual procedure,

$$\phi = \psi^{-1}\sum_{j=0}^{\infty}\phi_j\psi^j$$

resonances at $j = -1, 0, 1$.

The compatibility condition at $j = 0$ and $j = 1$ are satisfied, so the Painleve property is there. The Backlund map obtained under truncation

$$\phi = \phi_0/\psi + \phi_1 .$$ (3.3.17)

It may be noted that the most general form of the transformation is the homographic transformation. It is fairly easy to show that an equation of the form

$$A(\phi_t/\phi_x) + B(\{\phi, x\}) = 0 ,$$ (3.3.18)

where $A(\phi_t/\phi_x)$ and $B(\{\phi, x\})$ are constant coefficient multinomials in $\partial^k/\partial t^j \partial x^i \cdot (\phi_t/\phi_x)$ and $\partial^l/\partial t^m \cdot \partial x^n \cdot \{\phi, x\}$ respectively with,

$$j + i = k , \quad m + n = 1 , \quad \max(k) < \max(1) + 2$$

will have these properties.

An interesting variation is given by a new example

$$\frac{\phi_t}{\phi_x} + \frac{\partial}{\partial x}\{\phi, x\} = 0 .$$ (3.3.19)

The structure about the movable poles are of the form

$$\phi = \psi^{-1} \sum_{j=0}^{\infty} \phi_j \psi^j ,$$

with resonances at $j = -1, 0, 1, 2$.

It can be readily verified that the compatibility conditions are satisfied. On the other hand, if we set

$$\phi = v_1/v_2 , \quad v_{xx} = av$$

and

$$v_t = bv_x + cv ,$$

we find

$$v_{xx} = 1/2\omega v$$

$$v_t = \omega_x v_x - \frac{1}{2}\omega_{xx}v$$

(3.3.20)

and

$$\omega_t + \omega_{xxxx} = 2\omega\omega_x + \omega_x^2 .$$

(3.3.21)

But it is impossible to introduce a spectral parameter λ into (3.3.20) and (3.3.21) in such a manner that equation (3.3.19) will not depend on this parameter. This restricts the class of solution.

Now the singularities of (3.3.21) are of the form

$$\omega \simeq \omega_0/\psi^2 + ...$$

with resonances at

$$r = -1, \quad (7 \pm i\sqrt{11})/2, \quad 8 .$$

There exists a "Painleve type" expansion

$$\omega = \sum_{j=0}^{\infty} \omega_j \psi^{j-2} ,$$

(3.3.22)

where the compatibility at $j = 8$ is satisfied, but since the resonances $r = (7 \pm 11)/2$ are not included in (3.3.22), it represents only a special class. Furthermore, by substituting

$$V = \phi_{xx}/\phi_x ,$$

we find from (3.3.20)

$$V_t + \frac{\partial}{\partial x}\left(V_{xxx} - V_x^2 - V^2 V_x\right) = 0 .$$

The leading order analysis leads to

$$V \sim V_0 \phi^{-1} ,$$

where

$$V_0 = 3\phi_x , \qquad -2\phi_x .$$

Letting $V_0 = 3\phi_x$, we find resonances at $r = -1$, $(7 \pm i\sqrt{11})/2$. Again, the equation is Painleve at $r = 4$, but it allows the special form of solution

$$V = \sum_{j=0}^{\infty} V_j \phi^{j-1} .$$

For the other branch,

$$V_0 = -2\phi_x ,$$

the resonances occur at $r = -1, 2, 4, 5$. In this case, the expansion is

$$V = \sum_{j=0}^{\infty} V_j \phi^{j-1} ,$$

where (ϕ_1, V_2, V_4, V_5) are arbitrary.

By forming the Backlund transformation,

$$V = v_0/\phi + v_1 ,$$

we find that $v_0 = -2\phi_x$; $v_1 = \phi_{xx}/\phi_x$ and (V, V_1) also satisfy

$$\phi_t/\phi_x + \frac{\partial}{\partial x}\{\phi, x\} = 0 ,$$

which is just equation (3.3.19). So we find a curious phenomenon that equation (3.3.16) does not lead to a completely integrable system, but systems with special properties only, though it is homographic invariant.

3.4 Kowalevski's exponent, integrals of motion

Though it is taken for granted that the Painleve analysis is the test for complete integrability of a nonlinear system, a foolproof definition of complete integrability is still lacking. On the other hand, a different condition for a system to be completely integrable is that the system under condition permits a sufficient number of independent single-valued first integrals. In other words, the question that still haunts people is, for a given dynamical system, what can we state about its complete integrability in a finite procedure? The universality of the existence theorem on the solution of ordinary differential equations suggests that the integrability of dynamical systems cannot influence the local character of solution as long as the regular domain of solution is concerned. That is why integrability is usually discussed in connection with the global behavior of solutions. But the integrability of dynamic systems may influence a local character at a singularity of the solution. This important question was raised and studied in detail by Yoshida[51,52] in two lucid papers. Of course, his attention was only on systems that possess a special sort of invariance, the so called scaling invariance, which is a special case of conformal transformation.

Kowalevski's exponents

Consider the system of autonomous differential equations

$$\frac{dx_i}{dt} = F_i(x_1,\ldots,x_n); \quad i = 1,\ldots,n ,$$ (3.4.1)

where, $F_1,\ldots,F_n$ are rational functions of x_i. We say that the system is similarity invariant when there is a set of rational numbers $g_1,\ldots,g_n$ such that it is invariant under the similar transformation,

$$t \to \alpha^{-1}t; \quad x_1 \to \alpha^{g_1}x_1; \quad x_2 \to \alpha^{g_2}x_2\ldots$$

for a constant ϕ. In other words, we should have

$$F_i\left(\alpha^{g_1}x_1,\ldots,\alpha^{g_n}x_n\right) = \alpha^{g_i+1}F_i\left(x_1,\ldots,x_n\right) \quad i = 1,\ldots,n$$ (3.4.2)

simultaneously for arbitrary x and α. Differentiating with respect to α and setting $\alpha = 1$, we get

$$\sum g_i x_i \frac{\partial F_j}{\partial x_i}(x_1,\ldots,x_n) = \left(g_j + 1\right)F_j(x_1,\ldots,x_n) ,$$ (3.4.3)

which is a system of linear algebraic equations that determine the unknowns $g_1,...,g_n$ when F_i is known. So if

$$\det_{1<i,j<n}\left|x_j\frac{\partial F_i}{\partial x_j}(x_1,...,x_n)-\delta_{ij}F_i(x_1,...,x_n)\right|\neq 0 \; ,$$

the choice of $g_1,...,g_n$ is unique, if at least one of them exists. It is important to mention that for linear system no similarity transformation exists and as such, such an invariance is characteristic of a nonlinear one. Such a system has a special type of particular solution

$$x_1 = c_1 t^{-g_1},...,x_n = c_n t^{-g_n} \tag{3.4.4}$$

with constant c_i to be determined. Now from equation (3.4.2) we can deduce that

$$F_i\left(c_1 t^{-g_1},...,c_n t^{-g_n}\right)=t^{-g_{i-1}}F_i(c_1,...,c_n)$$

$i = 1,...,n$ by setting $x_1 = c_1,...,x_n = c_n$ and $1 \; \alpha = t^{-1}$. It in turn implies that the set $c_1,...,c_n$ are a set of solutions of the system of algebraic equation

$$F_i(c_1,...,c_n)=-g_1 c_i \; , \quad i=1,...,n \; .$$

Since the solutions are in general complex numbers, the particular solution noted above can be viewed as complex analytic function in the complex t-plane. We next consider the variational equation of (3.53) around the solution

$$\frac{dy_i}{dt}=\sum\frac{\partial F_i}{\partial x_j}\left(c_1 t^{-g_1},...,c_n t^{-g_n}\right)y_j \quad i=1,...,n \; . \tag{3.4.5}$$

Differentiating the scaling relations (3.54) and setting $\alpha = t^{-1}$, $x_i = c_i$, we have

$$t^{-g_j}\frac{\partial F_i}{\partial x_j}\left(c_1 t^{-g_1},...,c_n t^{-g_n}\right)=t^{-g_{i-1}}\frac{\partial F_i}{\partial x_j}(c_1,...,c_n) \; , \tag{3.4.6}$$

hence (3.57) can be rewritten as

$$\frac{dy_i}{dt} = \sum_{j=1}^{n} \frac{\partial F_i}{\partial x_j}(c_1,...,c_n)t^{+g_j-g_i-1}y_j \quad i=1,...,n. \tag{3.4.7}$$

It is then easy to observe that

$$y_1 = y_{10}t^{\rho-g_1},...,y_n = y_{n0}t^{\rho-g_n} . \tag{3.4.8}$$

Satisfy equation (3.4.7), when $y_0 = (y_{10},...,y_{n0})$ is an eigenvector of a $n \times n$ complex matrix K where

$$K_{ij} = \frac{\partial F_i}{\partial x_j}(c_1,...,c_n)+\delta_{ij}g_i . \tag{3.4.9}$$

The importance of the eigenvalues of K was first noted by Kowalevski and then by Liapunov, so we call

$$\det|\lambda\delta_{ij} - K_{ij}| \tag{3.4.10}$$

the Kowalevski determinant. The eigenvalues of K are called Kowalevski's exponent. For example, consider the Hamiltonian with two degrees of freedom

$$H = 1/2\left(p_1^2 + p_2^2\right) + q_1^2 q_2 + \frac{\varepsilon}{3}q_2^3 ,$$

where ε is a constant. Equations of motion are

$$dq_1/dt = p_1 ; \qquad dq_2/dt = p_2$$

$$dp_1/dt = -2q_1q_2 ; \qquad dp_2/dt = -q_1^2 - \varepsilon q_2^2 .$$

This system is invariant under

$$t \to \alpha^{-1}t, \quad q_1 \to \alpha^2 q_1, \quad q_2 \to \alpha^2 q_2, \quad p_1 \to \alpha^3 p_1, \quad p_2 \to \alpha^3 p_2 ,$$

then the determinant becomes

$$\det \begin{vmatrix} -p_1 & 0 & p_1 & 0 \\ 0 & -p_2 & 0 & p_2 \\ -2q_1q_2 & -2q_1q_2 & -2q_1q_2 & 0 \\ -2q_1^2 & -2\varepsilon q_2^2 & 0 & q_1^2 + \varepsilon q_2^2 \end{vmatrix} = -4q_1^3 q_2 p_1 p_2 \ ,$$

which is not identically 0. For this case, the particular solution can be written as

$$q_1 = c_1 t^{-2} \ , \quad q_2 = c_2 t^{-2} \ , \quad p_1 = c_3 t^{-3} \ , \quad p_2 = c_4 t^{-3}$$

along with

$$c_3 = -2c_1 \ , \quad c_4 = -2c_2$$

$$-2c_1 c_2 = -3c_3 \ , \quad -c_1^2 - \varepsilon c_2^2 = -3c_4 \ .$$

Eventually there are two different choices

(i) $c_1 = \pm\sqrt{9(2-\varepsilon)} \ ; \quad c_2 = -3$

(ii) $c_1 = 0 \ , \quad c_2 = -6/\varepsilon \cdot \varepsilon \neq 0 \ .$

Kowalevski's determinant is simplified to

$$K(\rho) = \det \begin{bmatrix} (\rho-2)(\rho-3)+2c_2 & 2c_1 \\ 2c_1 & (\rho-2)(\rho-3)+2\varepsilon c_2 \end{bmatrix}$$

which reduces to

(i) $K(\rho) = (\rho+1)(\rho-6)\{\rho^2 - 5\rho + 6(2-\varepsilon)\}$

(ii) $K(\rho) = (\rho+1)(\rho-6)\{\rho^2 - 5\rho + 6(1-2/\varepsilon)\} \ .$

Algebraic integrals and Brun's Lemma

Suppose our similarity invariant system admits a first integral, polynomial in x_i

$$H(x_1,...,x_n) = \text{const.}$$

Due to the invariance noted above, we should also have

$$H^1 = H\left(\alpha^{g_1} x_1,\ldots,\alpha^{g_n} x_n\right) = \text{constant}$$

as another integral. We say that a function $\phi(t,x_1,\ldots,x_n)$ is a weighted homogeneous function of degree M if

$$\phi\left(\alpha^{-1} t, \alpha^{g_1} x_1,\ldots,\alpha^{g_n} x_n\right) = \alpha^M \phi(t, x_1,\ldots,x_n) \ .$$

It is then possible to write the first integral as a sum of weighted homogeneous polynomials

$$H' = \sum \alpha^m H_m(x_1,\ldots,x_n) = \text{const.}$$

H_m stands for a weighted homogeneous polynomial of degree. Also a first integral

$$H(x_1,\ldots,x_n) = a \ , \tag{3.4.11}$$

is called an algebraic first integral when equation (3.4.11) can be rationalized into the form of an algebraic equation for a as

$$a^k + h_1(x_1,\ldots,x_n)a^{k-1}+\ldots+H_k(x_1,\ldots,x_n) = 0 \ . \tag{3.4.12}$$

Then Brun's Lemma states that if $F_1,\ldots,F_n$ are all rational functions of x, every coefficient $H_1,\ldots,H_k$ in (3.4.12) becomes first integral. We now go back to the basic equation of the dynamical system,

$$dx_i/dt = F_i(x_1,\ldots,x_n) \ . \tag{3.4.13}$$

For a solution $x_i = \phi_i(t)$, we define the variation

$$x_i = \phi_i(t) + y_i$$

so that the variational equation of (3.4.14) becomes

$$\frac{dy_i}{dt} = \sum_{j=1}^{n} \left(\frac{\partial F_i}{\partial x_j} \right) y_j \Bigg|_{x_i = \phi_i(t)} . \tag{3.4.15}$$

One can easily observe that

$$y_1 = \frac{d\phi_i(t)}{dt} = F_i(x_1,\dots,x_n) \Bigg|_{x_i = \phi_i} \qquad i = 1,\dots,n$$

is a solution set of (3.4.15). Now let $H(x_1,\dots,x_n) = $ const. be a first integral for (3.65). Then for any solution of the variational equation we have

$$\sum_{i=1}^{n} (\partial H/\partial x_i)_{x_i = \phi_i(t)} \quad y_i = \text{const.} \tag{3.4.16}$$

For simple proof of this assertion, compute

$$\frac{d}{dt} \sum_i \frac{\partial}{\partial x_i} y_i = \sum_i \sum_j \left\{ \frac{\partial H}{\partial x_i} \frac{\partial F_i}{\partial x_j} y_j + \frac{\partial^2 H}{\partial x_i \partial x_j} F_j y_i \right\}$$

$$= \sum_i y_i \left\{ \sum_j \frac{\partial H}{\partial x_j} \frac{\partial F_j}{\partial x_i} + \frac{\partial^2 H}{\partial x_i \partial x_j} F_j \right\}$$

$$= \sum_i y_i \frac{\partial}{\partial x_i} \left\{ \sum_j \frac{\partial H}{\partial x_j} F_j \right\} = 0 .$$

The last equality holds because

$$0 = \frac{dH}{dt} = \sum_j \frac{\partial H}{\partial x_j} F_j .$$

Next consider the case when the original system is given as a Hamiltonian system:

$$\frac{dq_i}{dt} = \frac{\partial H_0}{\partial p_i} ; \qquad \frac{dp_i}{dt} = -\frac{\partial H_0}{\partial q_1} .$$

Let a given solution be

$$q_{i\ i} = \phi_i(t) ; \quad p_i = \psi_i(t) ,$$

then the variations are

$$q_i = \phi_i(t) + \xi_i ; \quad p_i = \psi_i(t) + n_i ,$$

whence the variational equations are

$$\frac{d\xi_i}{dt} = \sum_k \frac{\partial^2 H}{\partial p_i \partial q_k} \xi_k + \sum_k \frac{\partial^2 H}{\partial p_i \partial p_k} n_k$$

$$\frac{dn_i}{dt} = -\sum_k \frac{\partial^2 H}{\partial q_i \partial q_k} \xi_k - \sum_k \frac{\partial^2 H}{\partial q_i \partial p_k} n_k .$$

Then the statement of Poincare Lemma goes as follows:

If $\phi(q_1,...,q_g, p_1,...,p_f) = $ const. is a first integral for the Hamiltonian system (3.4.13), then a solution of the variational equation is

$$\xi_i = \frac{\partial \Phi}{\partial p_i}\Big|_{\substack{q=\phi \\ p=\psi}} ; \quad n_i = -\frac{\partial \phi}{\partial q_i}\Big|_{\substack{q=\phi \\ p=\psi}} .$$

A simple demonstration of the above statement can be obtained by computing $d\xi_i/dt$:

$$\frac{d\xi_i}{dt} = \sum_k \left\{ \frac{\partial^2 \phi}{\partial p_i \partial q_k} \frac{\partial H_0}{\partial p_k} - \frac{\partial^2 \phi}{\partial p_i \partial p_k} \frac{\partial H_0}{\partial q_k} \right\}$$

$$= \sum_k \left\{ \frac{\partial^2 H_0}{\partial p_i \partial q_k} \frac{\partial \Phi}{\partial p_k} - \frac{\partial^2 H_0}{\partial p_i \partial p_k} \frac{\partial \Phi}{\partial q_k} \right\} + \frac{\partial}{\partial p_i} \sum_k \left\{ \frac{\partial \Phi}{\partial q_k} \frac{\partial H_0}{\partial p_k} - \frac{\partial \Phi}{\partial p_k} \frac{\partial H_0}{\partial q_k} \right\}$$

$$= \sum_k \frac{\partial^2 H_0}{\partial p_i \partial q_k} \xi_k + \sum_k \frac{\partial^2 H_0}{\partial p_i \partial p_k} n_k .$$

Similarly,

$$\frac{dn_i}{dt} = -\sum_k \frac{\partial^2 H_0}{\partial q_i \partial q_k} \xi_k - \sum_k \frac{\partial^2 H_0}{\partial q_i \partial p_k} n_k .$$

So by combining Brun's and Poincare's Lemma, one can assert that the existence of any algebraic integral is necessarily reduced to the existence of a weighted homogeneous rational first integral for similarity invariant system. The weighted degree appears as a kn Kowalevski's exponent under some condition. These results have been summarized in the main theorem of Yoshida's paper:

Let $\phi(x_1,\ldots,x_n) = \text{const.}$ be a weighted homogeneous first integral of weighted degree M for the similarity invariant system (3.4.13). Suppose the elements of the vector

$$\text{grad } \phi(x) = \left[\frac{\partial \phi}{\partial x_1}, \ldots, \frac{\partial \phi}{\partial x_n} \right]$$

are finite and not 0, then $M = \rho$ becomes a Kowalevski exponent. Different situations arise when the matrix K is not diagonalizable or when there may be more than one algebraic first integral. For the corresponding generalization, we refer to the famous papers by Yoshida. Note that for our previous example, the Hamiltonian system is the weighted homogeneous polynomial of weighted degree, so the Kowalevski determinant must have factor $(\rho+1)(\rho-6)$, which is true. Furthermore, it is known that special values of these exist for more first integrals independent of H.

$$\varepsilon = 1, \quad \phi_6 = p_1 p_2 + \frac{1}{3} q_1^3 + q_1 q_2^2$$

$$\varepsilon = 6; \quad \phi_8 = -4 p_1 (p_1 q_2 - p_2 q_1) + 4 q_1^2 q_2^2 + q_1^4$$

$$\varepsilon = 16; \quad \phi_{12} = \frac{1}{4} p_1^4 + q_1^2 q_2 p_1^2 - \frac{1}{3} q_1^3 p_1 p_2 - \frac{1}{18} q_1^6 - \frac{1}{3} q_1^4 q_2^2 .$$

The existence of these integrals requires that Kowalevski's exponents $\rho = M$ and $\rho = 5 - M$ for the choice of c_1 and c_2 for which the assumption of the theorem is satisfied. In fact, exponents other than -1 and 6 are

$$\varepsilon = 1, \quad \rho = 6, \text{ and } -1 \text{ for second choice of } c_1, c_2$$

$$\varepsilon = 6, \quad \rho = 8, \text{ and } -3 \text{ for first choice of } c_1, c_2$$

$$\varepsilon = 16, \quad \rho = 12, \text{ and } -7 \text{ for first choice of } c_1, c_2 .$$

Now consider another variant of the system discussed in the above example. This system is written as

$$q_{1t} = p_1, \quad q_{2t} = p_2, \quad p_{1t} = -3q_1^2 - \frac{1}{2}q_2^2, \quad p_{2t} = -q_1 q_2.$$

Governed by the Hamiltonian system,

$$H = 1/2\left(p_1^2 + p_2^2\right) - q_1^3 - \frac{1}{2}q_2^2 q_1,$$

it is again invariant under the transformation

$$t \to \alpha^{-1}t, \quad q_1 \to \alpha^2 q_1; \quad q_2 \to \alpha^2 q_2, \quad p_1 \to \alpha^3 p_1, \quad p_2 \to \alpha^3 p_2.$$

But this time the determinant is

$$\begin{vmatrix} -p_1 & 0 & p_1 & 0 \\ 0 & -p_2 & 0 & p_2 \\ -6q_1^2 & -q_2^2 & -3q_1^2 - \frac{1}{2}q_2^2 & 0 \\ -2q_1q_2 & -q_1q_2 & 0 & q_1q_2 \end{vmatrix} = -3p_1 p_2\left(q_1 q_2^3 + q_1^3 q_2\right).$$

Again, it is not identically 0. We can again search for solution in the form

$$q_1 = c_1 t^{-2}, \quad q_2 = c_2 t^{-2}, \quad p_1 = c_3 t^{-3}, \quad p_2 = c_4 t^{-3},$$

whence c_i satisfies a set of equations similar to that of the previous case and the full analysis can be repeated. The reason for mentioning this last example is that this is a variant of Henon-Heils system found to pass the full Painleve analysis by Chang, Tabor, and Weiss.

The case of PDE

Since Painleve analysis has been extended to the case of nonlinear Partial Differential Equations (nPDE) also, it will be very interesting to extend the analysis of Yoshida to the case of nPDE. One first notes that instead of first (algebraic) integrals we have the conservation law

$$\frac{\partial D[u]}{\partial t} + \frac{\partial J[u]}{\partial x} = 0 ,$$ (3.4.17)

where D is the density and J is the flow. Consider the case of KdV equation:

$$u_t - 6uu_x + u_{xxx} = 0 .$$ (3.4.18)

Setting $u = v_x$, the Lagrangion density of the equation

$$v_{xt} - 6v_x v_{xx} + v_{xxxx} = 0$$

is given as

$$L = -\frac{1}{2}v_x v_t + (v_x)^3 + \frac{1}{2}(v_{xx})^2 .$$

The corresponding Hamiltonian density is given by

$$H = v_t \frac{\partial L}{\partial v_t} - L = -(v_x)^3 - \frac{1}{2}(v_{xx})^2 .$$

We now observe that the equation is scale invariant under

$$x \to \alpha^{-1}x , \quad v \to \alpha v .$$

Consequently,

$$H(\alpha^2 v_x, \alpha^3 v_{xx}) = \alpha^6 H(v_x, v_{xx}) .$$

On the other hand, we already know that $r = 6$ is a resonance position. We can also cite the Liouville equation,

$$u_{tt} - u_{xx} + e^u = 0 .$$

Setting $v = e^u$, one gets

$$v(v_{tt} - v_{xx}) - v_t^2 + v_x^2 + v^3 = 0 .$$

The Hamiltonian system can be written as

$$H = \frac{1}{2}\left\{ \left(v_t\right)^2 v^{-2} + \left(v_x\right)^2 v^{-2} \right\} - v \ .$$

Now observe that equation (3.68) is scale invariant under $t \rightarrow \alpha^{-1}t$, $x \rightarrow \alpha^{-1}x$, $v \rightarrow \alpha^2 v$, whence

$$H\left(\alpha^2 v_1, \alpha^3 v_t, \alpha^3 v_x\right) = \alpha^2 H\left(v, v_t, v_x\right) \ .$$

Consequently, $r = 2$ is a resonance position, so we may observe that the position of the resonances in the Painleve test is intimately related to Kovalevski's exponent. Such a connection may not be very surprising because of the fact that many integrable nonlinear equations reduce to the original Painleve class of ODE under similar reduction.

3.5 Nonintegrable system and Painleve analysis

Until now we have seen that the Painleve analysis is very useful for the study of completely integrable PDEs and ODEs. The formalism yields the Backlund transformation, which in turn can generate various solutions by iteration, starting with different seed solutions. On the other hand, it has been observed that many nonlinear systems, such as the Kuramoto-Shivashinsky equation, and the Boussinesq-Schrodinger system, which were not known to be completely integrable, can also be studied by this technique. And in many cases, one can obtain some specialized solutions of such nonlinear systems. In fact, it has been seen that some chaotic phenomena can be modeled by the Kuramoto-Shivaashinsky system. So the Painleve test not only serves to analyze integrable system, but it can also be used effectively to generate solutions of nonintegrable problems that cannot have a Lax pair. The following discusses in detail some important examples that will illustrate the methodology in its fullest detail. Another important point worth mentioning is the relation between the Painleve analysis and Hirota's method of bilinearisation. It has been found that many nonlinear PDEs may not be completely integrable but can be effectively analyzed by Hirota's method, and it may be possible to construct multisoliton solutions. As such, Hirota's technique can also be effectively combined to study these nonintegrable PDEs.

Capillary-gravity waves

Let us begin with the equations of capillary gravity waves initially deduced by Kawahara *et al.* and later solved by Ma for the *N*-soliton solution. Actually, Ma obtained the

N-soliton solution with the help of Hirota's technique because no Lax pair was known for it. The equations under consideration can be written as

$$iE_t + E_{xx} = nE$$

$$-iG_t + G_{xx} = nG \qquad (3.5.1)$$

$$n_t - 6nn_x + n_{xxx} = -(EG)_x$$

Following the procedure explained in Chapter 2 we can set

$$E = \phi^p \sum a_j \phi^j(x,t)$$

$$G = \phi^q \sum b_j \phi^j(x,t)$$

$$n = \phi^s \sum c_j \phi^j(x,t)$$

But instead we pursue a different approach. A similar but more general form of equation occurs in plasma, which is written as

$$i\left(\frac{\partial E}{\partial t} + V_g \frac{\partial E}{\partial x}\right) + \frac{D_0}{2} - \frac{\partial^2 E}{\partial x^2} = \mu \omega N E$$

$$\frac{\partial N}{\partial t} + V_M \frac{\partial N}{\partial x} + \frac{\theta^2}{2V_M} \frac{\partial^3 N}{\partial x^3} + \frac{a^2}{V_M} N \frac{\partial N}{\partial x} = -n^2 \frac{\partial}{\partial x}\left(\frac{|E|^2}{16\pi n_0 Te}\right), \qquad (3.5.2)$$

where E stands for the electric field and N is the density inside the plasma. V_g is the group velocity, and V_M is the magnetoacoustic speed. We consider the stationary form of these equations, which turn out to be

$$D_0 \frac{d^2 E}{d\xi^2} = \lambda E + b^2 NE \qquad (3.5.3a)$$

$$\theta^2 \frac{d^2 N}{d\xi^2} = fN - a^2 N^2 - n^2 \frac{E^2}{16\pi n_0 Te}, \qquad (3.5.3b)$$

where $f = 2V_M(M - V_M)$ and $= 2\delta + (M^2 - V_g^2)/D_0$ and $\xi = x - Mt$, M being the Mach number. Equations (3.5.3a) and (3.5.3b) can be reduced to the standard form by normalizing N and E^2, respectively. For negative dispersion, these reduced systems become

$$\frac{d^2 E}{d^2} = -AE - 2DEN \qquad (3.5.4a)$$

$$\frac{d^2 N}{d^2} = -BN - CN^2 - DE^2 \ , \qquad (3.5.4b)$$

where

$$A = \frac{\lambda}{|D_0|} \ ; \quad B = -\frac{f}{\theta^2} \ ; \quad C = \frac{2a^2 n^2 |D_0|}{b^2 \theta^4} \ ; \quad D = \frac{n^2}{\theta^2} \ .$$

It is not difficult to demonstrate that this equation set has the Hamiltonian system

$$H_+ = 1/2\left(\pi_E^2 + \pi_N^2\right) + 1/2\left(AE^2 + BN^2\right) + \left(\tfrac{1}{3}CN^3 + DNE^2\right) ,$$

where π_E, π_N are the canonical momenta.

Similarly, in cases of positive dispersion, the coupled set becomes

$$\frac{d^2 E}{d^2} = AE - 2DEN \qquad (3.5.5a)$$

$$\frac{d^2 N}{d^2} = -BN + CN^2 + DE^2 \ , \qquad (3.5.5b)$$

derivable from

$$H_- = 1/2\left(\pi_E^2 - \pi_N^2\right) - 1/2\left(AE^2 + BN^2\right) + \left(\tfrac{1}{3}CN^3 + DNE^2\right) .$$

For the integrability of the system (3.5.4a,b) or (3.5.5a,b), we need to have second integral of motion I, which is in involution with the Hamiltonian system, that is $\{H_\pm, I\}$ must vanish. We can observe that

$$I = 4E\left\{ p\pi_E\pi_N + AEN + \frac{p}{3}DE^3 + DN^2E \right\} - 4n\pi_E^2 + \frac{p}{3}DE^4 + \frac{4A - pB}{D}\left(\pi_E^2 + AE^2 \right)$$

when A, B are arbitrary but $C = 6pD$, and

$$I = p\pi_E\pi_N + AEN + \frac{p}{3}DE^3 + DN^2E$$

for the situation $B = pA$, $C = pD$. Lastly,

$$I = \pi_E^4 + 2(A + 2DN)E^2\pi_E^2 + A^2E^4 + \frac{2p}{9}D^2E^6$$

$$- \frac{4}{3}DE^3\left(p\pi_E\pi_N + AEN + \frac{p}{3}DE^3 + DN^2E \right)$$

when $B = 16pA$, $C = 16pD$, $p = \pm 1$.

We can now eliminate one of the variables, say E, from the coupled system and obtain

$$\frac{d^4N}{d\xi^2} + [(4A + pB) + 2(pC + 4D)N]\frac{d^2N}{d\xi^2} + 2(pC - D)$$

$$\times \left(\frac{dN}{d\xi}\right)^2 + \left[(4pAB)N + 2p(2AC + 3BD)N^2 + \left(\frac{20}{3}pCD\right)N^3\right]$$

$$+ 4pDH = 0 .\tag{3.5.6}$$

A special situation where the linear terms A, B are absent deserves special attention. If $A = B = 0$, we get

$$\left[N_{\xi\xi\xi\xi} + 2(pC + 4D)NN_{\xi\xi} + 2(pC - D)\left(\sum N_\xi\right)^2 + \frac{20}{3}pCDN^3 \right] + 4pDH = 0 .$$

On the other hand, there are quite a few fifth-order nonlinear equations whose scale reduced form is given by (3.5.6) above. These are

• Lax's fifth order equation

$$u_t = \left[u_{yyyy} + \lambda_1 uu_{yy} + 1/2(\lambda_2 - \lambda_1)(u_y)^2 + \lambda \frac{3}{3}u^3 \right]$$

- Swada-Kotera equation

$$u_t = \left[u_{yyyy} + 5uu_{yy} + \frac{5}{3}u^3 \right]_y$$

- Kaup Kuperschmidt equation

$$u_t = \left[u_{yyyy} + 10uu_{yy} + \frac{15}{2}\left(u_y\right)^2 + \frac{20}{3}u_3 \right]_y .$$

It can be ascertained that those stationary equations are integrable also. But these approaches are not foolproof, and integrability of the original coupled set is not guaranteed. So by the heuristic approach (which is a simplified version of AKS), one can obtain a lot of useful information, despite the fact that no Lax pair is known for this set.

Coupled Schrodinger and Boussinesq system
The equations under consideration read[43,44]

$$\left(\frac{\partial^2}{\partial t^2} - \frac{\partial^2}{\partial x^2} - \frac{\sigma}{3}\frac{\partial^4}{\partial x^4} \right)u - \sigma\frac{\partial^2}{\partial x^2}\left(u^2\right) = \frac{\partial^2}{\partial x^2}\left(|\psi|^2\right)$$

$$\left(i\frac{\partial}{\partial t} + \frac{\partial^2}{\partial x^2} + \lambda + u \right)\psi = 0 .$$

One can also rewrite this set as

$$\left(\frac{\partial^2}{\partial t^2} - \frac{\partial^2}{\partial x^2} - \frac{\sigma}{3}\frac{\partial^4}{\partial x^4} \right)u - \sigma\frac{\partial^2}{\partial x^2}\left(u^2\right) = \frac{\partial^2}{\partial x^2}(\chi\psi)$$

$$\left(i\frac{\partial}{\partial t} + \frac{2}{\partial x^2} + \lambda + u \right)\psi = 0 \tag{3.5.7}$$

$$\left(-\frac{\partial}{\partial t} + \frac{\partial^2}{\partial x^2} + \lambda + u \right)X = 0 .$$

The Singularity analysis starts by adopting the ansatz

$$u = \sum_{j=0}^{\infty} u_j(x,t)\phi^{\alpha+j}(x,t) \ ,$$

$$\psi = \sum_{j=0}^{\infty} \psi_j(x,t)\phi^{\beta+j}(x,t) \ ,$$

$$\chi = \sum_{j=0}^{\infty} X_j(xt)\phi^{\gamma+j}(x,t) \ .$$

The singularity manifold is given as $\phi = 0$. The leading order analysis leads to two alternatives,

(a) $\alpha = -2, \ \beta = \gamma = -1$

$u_0 = -2\phi_x^2 \cdot (\psi_0, \chi_0)$ arbitrary.

Matching terms are

$$\sigma/3u_{xxxx} + \sigma\left(u^2\right)_{xx} = 0$$

$$\psi_{xx} + u = 0 , \qquad X_{xx} + uX = 0$$

(b) $\alpha = -2, \ \beta = \gamma = -2, \ u_0 = -6\phi_x^2$

$\chi_0\psi_0 = -24\sigma\phi_x^4$

Dominant terms are

$$\frac{\sigma}{3}u_{xxxx} + \sigma\left(u^2\right)_{xx} = (\chi\psi)_{xx}$$

$$\psi_{xx} + u\psi = 0$$

$$X_{xx} + u\chi = 0 \ .$$

It may be mentioned that in the determination of the leading order coefficients in case (a) the field ψ and χ become decoupled, but not in case (b). Such a situation also occurs in the Hirota-Satsuma case.

Resonance determination

To start with, consider the situation (a). We set

$$u = \sum_{j=0}^{\infty} u_j \phi^{j-2} \ ,$$

$$\chi = \sum_{0}^{\infty} \chi_j \phi^{j-1} \ ,$$

$$\psi = \sum_{0}^{\infty} X_j \phi^{j-1} \ ,$$

in equation (3.5.7) and equate the same power of ϕ to arrive at the following recurs relation,

$$\begin{pmatrix} N\phi_x^4 & 0 & 0 \\ \psi_0 & M\phi_x^2 & 0 \\ \chi_0 & 0 & M\phi_x^2 \end{pmatrix} \begin{pmatrix} u_m \\ \psi_m \\ \chi_m \end{pmatrix} = \text{Other terms with } u_j, \chi; \ \psi_j \text{ when } j < m \quad (3.5.8)$$

the matrix on the left hand side is known as the system matrix denoted by T. In this expression for T, we have used

$$N = \sigma / 3(-2 + m)(-3 + m)(-4 + m)(-5 + m) + 4\sigma(-4 + m)(-5 + m)$$

$$M = (-1 + m)(-2 + m) - 2 \ .$$

Now resonances are determined by demanding that the determinant of the system matrix is equal to 0. So if we set

$$\det T = 0 \quad \text{for} \quad m = r \ ,$$

we get $r^2(r + 1)(r - 3)^2(r - 4)(r - 5)(r - 6) = 0$, whence resonances are at $r = 0, \ 0, \ -1, \ 3,$ 3, 4, 5, 6.

On the other hand, for the branch (6), the matrix T is written as

$$= \begin{pmatrix} N_1\phi_c^4 & L_1\phi_x^2\chi_0 & L_1\phi_x^2\psi_0 \\ \psi_0 & M_1\phi_x^2 & 0 \\ \chi_0 & 0 & M_1\phi_x^2 \end{pmatrix}$$

where

$$N_1 = \sigma / 3[-(-2+m)(-3+m)(-4+m)(-5+m)+36(-4+m)(-5+m)]$$

$$L_1 = -(-4+m)(-5+m)$$

$$M_1 = (-2+m)(-3+m)-6 \ .$$

We also have the relation similar to (3.5.8),

$$T \begin{pmatrix} u_m \\ m \\ m \end{pmatrix} = (\text{terms involving } u_j, \ \psi_j, \ \chi_j \text{ with } j < m).$$

From the condition that $\det T = 0$ we get $r = 0, -1, -3, 4, 5, 5, 6, 8$.

It is interesting to observe that in both cases the resonances are integers, but in the following it will turn out that only the first branch passes the test. There is another important aspect in these resonances – the existence of negative ones. At first sight it appears that these do not contribute to the existence of arbitrary function in the expansion. Significance of such negative resonances have been studied by Steeb and Louw and also by Fordy and Pickering. These will be discussed in Section 3.7. At this stage we can point out a very simple application of the relation between the algebraic constants of motion and the resonance position. Consider equation (3.5.7) with most singular term

$$1/3u_{xxxx} + \left(u^2\right)_{xx} = 0 \ .$$

This is scale invariant under $x \to \varepsilon^{-1}x, \ u \to \varepsilon.$

Since $\partial/\partial x \left[1/3u_{xxx} + \left(u^2\right)_x \right] = 0 = \partial I / \partial X$, we find, $I\left(\varepsilon^2 u, \ \varepsilon^3 u_x ...\right) = \varepsilon^5 I\left(u, \ u_x ...\right).$

Thus $r = 5$ must be a resonance, which has been obtained from the system matrix. The next important criterion for the Painleve analysis is the compatibility of the coefficients.

Arbitrariness of coefficients

If we write out equation (3.5.8) in full, we get

$$T \begin{pmatrix} u_m \\ m \\ m \end{pmatrix} = \begin{pmatrix} A \\ B \\ C \end{pmatrix}$$

$$A = -u_{(m-4)tt} + 2u_{(m-3)t}(m-5)_{f_t}$$
$$-u_{m-2}(m-4)\cdot(m-5)f_t^2 + u_{m-3}(m-5)_f + u_{(m-2)}(m-4)(m-5)$$
$$+\sigma\sum_{s=1}^{m-1}u_{m-s}u_s(m-4)(m-5) + \sum_{s=0}^{m-2}\psi_{m-s-2}(m-4)(m-5)$$

$$B = -i\psi_{(m-2)t} + i\psi_{(m-1)}(m-2)_{f_t} - \lambda\psi_{m-2} - \sum_{s=1}^{m-1}u_{m-s}\psi_s$$

$$C = i\chi_{(m-2)t} - i\chi_{(m-1)}(m-2)_{f_t} - \lambda\chi_{m-2} - \sum_{s=1}^{m-1}u_{m-s}\chi_s \ .$$

For the other branch,

$$s\begin{pmatrix} u_m \\ \psi_m \\ \chi_m \end{pmatrix} = \begin{pmatrix} D \\ E \\ F \end{pmatrix},$$

where

$$E = -i\psi_{(m-2)t} + i\psi_{(m-1)}(m-3)f_t - \lambda\psi_{(m-2)} - \sum_{s=1}^{m-1}u_{m-s}\psi_s$$

$$F = +i\chi_{(m-2)t} - i\chi_{m-1}(m-3)_{f_t} - \lambda\chi_{m-2} - \sum_{s=1}^{m-1}u_{m-s}\chi_s \ .$$

On the above computation, the Kruskal prescription, $\phi = x - f(t)$, has been used.

Branch (a)
$m = -1$, corresponds to arbitrary $f(t)$
$m = 0, 0$, corresponds to arbitrary $\lambda_0, \chi_0 : u_0 = -2$
$m = 1$, $u_1 = 0$

$$\psi_1 = i/2f_t\psi_0, \qquad \chi_1 = -i/2f_t\chi_0$$

$m = 2$, $u_2 = 1/4\sigma\left[2f_t^2 + (\psi_0\chi_0 - 2)\right]$

$$\psi_2 = \psi_0/8\sigma\left[2f_t^2 + (\psi_0\chi_0 - 2) + 4\lambda\sigma\right] + i/2\psi_{0t}$$

$$\chi_2 = \chi_0/8\sigma\left[2f_t^2 + (\psi_0\chi_0 - 2) + 4\lambda\sigma\right] - i/2\chi_{0t}$$

$m = 3$, $u_3 = 1/2\sigma f_{tt}$

$$u_3\psi_0 + i\psi_{1t} - i\psi_2 f_t + \lambda\psi_1 + u_2\psi_1 = 0$$
$$u_3\chi_0 - i\chi_{1t} + i\chi_2 f_t + \lambda\chi_1 + u_2\chi_1 = 0 \ .$$

A simple calculation leads to the fact that only for $\sigma = 1$ do we have ψ_3, χ_3 arbitrary.

$m = 4$. This leads to only two equations instead of three. We get

$$\psi_4 = -1/4\left[i\psi_{2t} - 2i\psi_3 f_t + \lambda\psi_2 + u_4\psi_0 + u_3\psi_1 + u_2\psi_2\right]$$
$$\chi_4 = -1/4\left[-i\chi_{2t} + 2i\chi_3 f_t + \lambda\chi_2 + u_4\chi_0 + u_3\chi_1 + u_2\chi_2\right]$$
u_4 = arbitrary.

$m = 5$. Here also we have a similar situation:

$$\psi_5 = -1/10\left(i\psi_{3t} - 3i\psi_4 f_t + \lambda\psi_3 + u_5\psi_0 + u_4\psi_1 + u_3\psi_2 + u_2\psi_3\right)$$
$$\chi_5 = -1/10\left(-i\chi_{3t} + 3i\chi_4 f_t + \lambda\chi_3 + u_5\chi_0 + u_4\chi_1 + u_3\chi_2 + u_2\chi_3\right)$$
u_5 = arbitrary.

$m = 6$. Here also we are led to the condition $\sigma = 1$ and

$$\psi_6 = -1/18\left(i\psi_{4t} - 4i\psi_5 f_t + \lambda\psi_4 + u_6\psi_0 + u_5\psi_1 + u_4\psi_2 + u_3\psi_3 + u_2\psi_4\right)$$
$$\chi_6 = -1/18\left(-i\chi_{4t} + 4i\chi_5 f_t + \lambda\chi_4 + u_6\chi_0 + u_5\chi_1 + u_4\chi_2 + u_3\chi_3 + u_2\chi_4\right)$$
u_6 = arbitrary.

From the above analysis, we infer that for $\sigma = 1$, $f(t)$, u_4, u_5, u_6, ψ_0, χ_0, ψ_3, χ_3 remain arbitrary satisfying the Cauchy-Kovalevski theorem. The number of arbitrary functions is same as the number of resonances, so we can conclude that this branch passes the Painleve test in the sense of Weiss et $al.$ for $\sigma = 1$.

Branch (b)
The results for a similar analysis in the branch (b) are as follows.
Here $m = -1$ also leads to the arbitrariness of $f(t)$.

For $m = 0$, we get $u_0 = -6$, $\chi_0\psi_0 = -24\sigma$, which corresponds to the arbitrariness of ??.

For $m = 1$, $u_1 = 0$, $\psi_1 = 1/2i\psi_0 f_t$, $\chi_1 = -1/2i\chi_0 f_t$.

For $m = 2$, this yields

$$u_2 = \frac{r}{60\sigma} + 1/20\sigma\left[(6-2\sigma)f_t^2 - 8\lambda\sigma - 2\right]$$

$$\psi_2 = i/6\psi_{0t} + \psi_0\left\{r/360\sigma + 1/120\sigma\left[(\sigma-12\sigma)f_t^2 + 12\lambda\sigma - 6\right]\right\}$$

$$\chi_2 = -i/6 \cdot \chi_{0t} + \chi_0\left\{r/360\sigma + 1/120\sigma\left[(6-12\sigma)f_t^2 + 12\lambda\sigma - 6\right]\right\} ,$$

where $r = i\chi_0\psi_{0t}$.

$m = 3$ gives

$$u_3 = [(2\sigma+3)/10\sigma]f_{tt}$$

$$\psi_3 = -(\sigma-1)/20\sigma\psi_0 f_{tt} - f_t/12\psi_{0t}$$
$$+ if_t/12\left\{r/60\sigma + 1/20\sigma\left[(6-2\sigma)f_t^2 + 12\lambda\sigma - \sigma\right]\right\}\psi_0$$

$$\chi_3 = -(\sigma-1)/20\sigma\psi_0 f_t - f_t/12 \cdot \chi_{0t}$$
$$- if_t/12\left\{r/60\sigma + 1/20\sigma\left[(\sigma-2\sigma)f_t^2 + 12\lambda\sigma - \sigma\right]\right\}\chi_0 .$$

For $m = 4$, we get only two equations:

$$\psi_4 = 1/4\left(i\psi_{2t} - i\psi_3 f_t + \lambda\psi_2 + u_3\psi_1 + u_2\psi_2 + \psi_0 u_4\right)$$

$$\chi_4 = 1/4\left(-i\chi_{2t} + i\chi_3 f_t + \lambda\chi_2 + u_3\chi_1 + u_2\chi_2 + \chi_0 u_4\right)$$

$$u_4 = \text{arbitrary.}$$

$m = 5$ leads to the conclusion that u_5 and ψ_5, χ_5 are arbitrary because

$$\psi_0 u_5 = -\left(i\psi_{3t} - 2i\psi_4 f_t + \lambda\psi_3 + u_4\psi_1 + u_3\psi_2 + u_2\psi_3\right)$$

$$\chi_0 u_5 = -\left(-i\chi_{3t} + 2i\chi_4 f_t + \lambda\chi_3 + u_4\chi_1 + u_3\chi_2 + u_2\chi_3\right) .$$

To check the consistency between these equations, we eliminate u_5. Then the previous results lead to

$$i(\sigma-1)\left[6f_{ttt}+5f_t^3\right]=0 \; .$$

So if $\sigma=1$, f remains arbitrary, otherwise f is determined. So for $\sigma=1$, we can proceed to higher resonances, $m=6,7,8$. It is interesting to note that even when $\sigma\ne 1$, that is when

$$6f_{ttt}+5f_t^3=0 \; , \qquad\qquad\qquad (3.5.9)$$

though $f(t)$ is determined, yet equation (3.5.9) itself passes the Painleve test. Now, for $\sigma=1$, $m=6$ using $u_5=0$, $u_6=$ arbitrary and

$$\psi_6=-\psi_0 u_0/6-1/6\left(i\psi_{4t}-3i\psi_5 f_t+\lambda\psi_4+u_4\psi_2+u_3\psi_3+u_2\psi_4\right)$$

$$\chi_6=-\chi_0 u_0/6-1/6\left(-i\chi_{4t}+3i\chi_5 f_t+\lambda\chi_4+u_4\chi_2+u_3\chi_3+u_2\chi_4\right) \; ,$$

we get along with the compatibility condition

$$-if_t\left(\chi_0\psi_5-\psi_0\chi_5\right)+1/3\left(\chi_0\psi_{4t}-\psi_0\chi_{4t}\right)+\lambda/3$$

$$\times\left(\chi_0\psi_4+\psi_0\chi_4\right)+u_4/3\left(\chi_0\psi_2+\psi_0\chi_2\right)+u_3/3\left(\chi_0\psi_3+\psi_0\chi_3\right)$$

$$+u_2/3\left(\chi_0\psi_4+\psi_0\chi_4\right)$$

$$=-u_{2tt}+2u_{3t}+\left(-2f_t^2+2+4\sigma u_2\right)u_4+u_3 f_{tt}+2\sigma u_3^2$$

$$+2\left(\psi_5\chi_1+\psi_1\chi_5\right)+2\left(\psi_4\chi_2+\psi_2\chi_4\right)+2\psi_3\chi_3 \; ,$$

which is satisfied by our previous results. For $m=7$ $(\sigma=1,\ u_5=0)$,

$$u_7=1/32\left\{-\left[u_{3tt}-4u_{4t}f_t-2u_4 f_{tt}-12u_3 u_4\right]+6\left(\psi_7\chi_0+\psi_0\chi_7\right)\right.$$

$$\left.+6\left(\psi_6\chi_1+\psi_1\chi_6\right)+6\left(\psi_5\chi_2+\psi_2\chi_5\right)+6\left(\psi_4\chi_3+\psi_3\chi_4\right)\right\}$$

$$\psi_7=-1/14\left(-i\psi_{5t}-4i\psi_6 f_t+\lambda\psi_5+u_6\psi_1+u_4\psi_3+u_3\psi_4+u_2\psi_5+\psi_0 u_7\right)$$

$$\chi_7=-1/14\left(-i\chi_{5t}+4i\chi_5 f_t+\lambda\chi_5+u_6\chi_1+u_4\chi_3+u_3\chi_4+u_2\chi_5+\chi_0 u_7\right) \; .$$

For $m=8$ $(\sigma=1,\ u_5=0)$,

$$u_8=\text{ arbitrary},$$

and we also have

$$\psi_8 = -1/24\left[i\psi_{0t} - 5i\psi_7 f_t + \lambda\psi_6 + u_7\psi_1 + (u_6\psi_2 + u_2\psi_6) + u_3\psi_5 + u_4\psi_4 + u_8\psi_0\right]$$

$$\chi_8 = -1/24\left[-i\chi_{0t} + 5i\chi_7 f_t + \lambda\chi_6 + u_7\chi_1 + (u_6\chi_2 + u_2\chi_0) + u_3\chi_5 + u_4\chi_4 + u_8\chi_0\right] ,$$

along with the compatibility condition

$$u_{4tt} + 12u_6 f_t^2 - 12u_6 - 12\left(12u_6 u_2 + u_4^2\right)$$
$$= -1/2\left\{i(\chi_0\psi_{6t} - \psi_0\chi_{6t}) - 5i(\chi_0\psi_7 - \psi_0\chi_7)f_t\right.$$
$$+(\lambda + u_2)(\chi_0\psi_6 + \psi_0\chi_6) + u_6(\chi_0\psi_2 + \psi_0\chi_2)$$
$$+u_4(\chi_0\psi_4 + \psi_0\chi_4) + u_3(\chi_0\psi_5 + \psi_0\chi_5) + 12(\psi_7\chi_1 + \psi_1\chi_7)$$
$$+12(\psi_5\chi_3 + \psi_3\chi_5) + 12\psi_4\chi_4 .$$

This equation can be seen to be satisfied by the previous results.

So we have checked all the conditions arising at the various resonance position except at $r = -3$. So actually we have eight resonances and seven arbitrary functions because $r = -3$ does not contribute to the summation. Unless the situation with such a negative resonance position is clarified (see Section 3.7), we cannot conclude that this branch passes the Painleve test.

Truncation of expansion
One of the most important aspects of the Painleve analysis is that in many situations it is possible to truncate the expansion over the singular manifold at finite number of terms. But after truncation, one has an overdetermined set of equations whose consistency is not at all obvious. Let us start from above equations (of branch) and set

$$u_3 = u_4,\ldots,= 0 , \qquad \psi_i = \chi_i = 0 \quad \text{for} \quad j > 2$$

leading to (note we always set $\sigma = 1$)

$$j = 1, \quad u_1 = 2\psi_{xx}$$

$$j = 2, \quad -12\phi_x^2 u_2 + 6\phi_t^2 - 8\phi_x\phi_{xxx} + 6\phi_{xx}^2 + 3\psi_0\chi_0 - 6\phi_x^2 = 0 \qquad (3.5.8)$$

$j = 3$, $24\phi_x\phi_t\phi_{xt} + 6\phi_{xx}\phi_t^2 + 6\phi_x^2\phi_{tt} - 36\phi_x^2\phi_{xx} - \left[18\phi_x^2\phi_{xxx}\right.$

$\left. -6\phi_{xx}^3 + 72\phi_x^2\phi_{xx}u_2 + 24\phi_x^3u_{2x}\right] + 9(\psi_0\chi_0)_x\phi_x = 0$ (3.5.9)

$j = 4$, $-12\phi_{xt}^2 - 12\phi_x\phi_{xtt} - 12\phi_t\phi_{xxt} - 6\phi_{xx}\phi_{tt} + 18\phi_{xx}^2$

$+24\phi_x\phi_{xxx} - \left[4\phi_{xxx}^2 - 6\phi_{xx}\phi_{xxxx} - 12\phi_x\phi_{xxxxx}\right.$

$-12\phi_x^2u_{2xx} - 36\phi_{xx}^2u_2 - 48u_2\cdot\phi_x\phi_{xxx} - 72\phi_x u_{2x}\phi_{xx}\right]$

$= (3\psi_0\chi_0)_{xx} - 6\phi_x(\psi_0\chi_1 + \psi_1\chi_0)_x - 3(\psi_0\chi_1 + \psi_1\chi_0)\phi_{xx}$ (3.5.10)

$j = 5$, $6\phi_{xtt} - 6\phi_{xxxx} - \left(2\phi_{xxxxx} + 12\phi_{xx}u_{2xx} + 12\phi_{xxxx}u_2 + 32u_{2x}\phi_{xxx}\right)$

$= 3(\psi_1\chi_0 + \chi_1\psi_0)_{xx}$.   (3.5.11)

Lastly, at $j = 6$

$$u_{2tt} - u_{2xx} - 1/3u_{2xxxx} - \left(u_2^2\right)_{xx} = (\psi_1\chi_1)_{xx} .$$

Similarly, if we start from the set of equation (of branch b), we get the following:

At $j = 1$,

$-i\psi_0\phi_t - 2\psi_{0x}\phi_x + \psi_0\phi_{xx} - 2\phi_x^2\psi_1 = 0$

$i\chi_0\phi_t - 2\chi_{0x}\phi_x + \chi_0\phi_{xx} - 2\phi_x^2\chi_1 = 0$ (3.5.12)

At $j = 2$,

$i\psi_{0t} + \psi_{0xx} + \lambda\psi_0 + 2\phi_{xx}\psi_1 + u_2\psi_0 = 0$

$-i\chi_{0t} + \chi_{0xx} + \lambda\chi_0 + 2\phi_{xx}\chi_1 + u_2\chi_0 = 0$ (3.5.13)

At $j = 3$,

$i\psi_{1t} + \psi_{1xx} + \lambda\psi_1 + u_2\psi_1 = 0$

$-i\chi_{1t} + \chi_{1xx} + \lambda\chi_1 + u_2\chi_1 = 0$ (3.5.14)

One should also note the following relation that has been repeatedly used in our subsequent calculation,

$\phi_x^2(\chi_0\psi_1 + \psi_0\chi_1) = (\psi_0\chi_0)\phi_{xx} - (\psi_0\chi_0)_x\phi_x$.   (3.5.15)

Now after truncation,

$$u = 2(\log \phi)_{xx} + u_2$$
$$\psi = \psi_0/\phi + \psi_1 \qquad\qquad (3.5.16)$$
$$\chi = \chi_0/\phi + \chi_1 ,$$

when u_2, ψ_1, χ_1 also satisfy the same set (3.5.7). But this system can be true if and only if the system of conditions written above is satisfied identically.

Compatibility of the overdetermined set
We observe that

$$(3.5.13) \times \chi_0 + (3.5.14) \times \psi_0$$
$$= -2(\psi_0\chi_0)_x \phi_x + 2(\psi_0\chi_0)\phi_{xx} - 2\phi_x^2(\psi_0\chi_1 + \chi_0\psi_1) = 0$$

or

$$(\psi_0\chi_0)\phi_{xx} - (\psi_0\chi_0)_x \phi_x = \phi_x^2(\psi_0\chi_1 + \chi_0\psi_1) .$$

Also from equations (3.5.8) and (3.5.9), we get

$$\phi_x^2\left[6\phi_{tt} - 6\phi_{xx} - 2\phi_{xxxx} - 12\phi_{xx}u_2 - 3(\chi_0\psi_1 + \psi_0\chi_1)\right] = 0 .$$

On the other hand, from (3.5.8) we get

$$u_2 = 1/12\left[6f_t^2 - 6 + 3(\psi_0\chi_0)\right] .$$

Also, from (3.5.13) and (3.5.14),

$$\psi_1 = 1/2\left[i\psi_0 f_t - 2\psi_{0x}\right], \quad \chi_1 = 1/2\left[-i\chi_0 f_t - 2\chi_{0x}\right] ,$$

along with

$$\chi_0\psi_1 + \psi_0\chi_1 = -(\psi_0\chi_0)_x .$$

We now substitute all of these in

$$i\psi_{it} + \psi_{1xx} + \lambda\psi_1 + u_2\psi_1$$

to observe that this expression vanishes identically if and only if $u_{2x} - 1/2 f_{tt} = 0$, which is a simple consequence of (3.5.9). So the set of equations are mutually compatible. A similar analysis can also be done for the other equations of the system.

Hirota's approach

A general proof of the truncation and thereby justification of the Backlund transformation can be made if we take recourse to the bilinear technique of Hirota. We set

$$u = (2\log f)_{xx} = D_x^2 \, ff/f^2 , \quad \psi = g/f, \quad \chi = h/f ,$$

then the basic equations can be rewritten

$$\left(D_t^2 - D_x^2 - \sigma/3D_x^4 + \rho\right)f \cdot f = gh \tag{3.5.17}$$

$$\left(iD_t + D_x^2 + \lambda\right)g \cdot f = 0$$

$$\left(-iD_t + D_x^2 + \lambda\right)h \cdot f = 0 . \tag{3.5.18}$$

The observation that the Backlund transformation can be written in the following way

$$f^{(n)} = \phi_{(n-1)} f^{(n-1)}$$

$$g^{(n)} = \psi_0 f^{(n-1)} + \phi_{(n-1)} g^{(n-1)} \tag{3.5.19a}$$

$$h^{(n)} = \chi_0 f^{(n-1)} + \phi_{(n-1)} h^{(n-1)}$$

prompts us to combine the methods of Weiss and Hirota. Let us now assume that $\left(f^n, g^n, h^n\right)$ satisfies the set (3.5.17) and (3.5.18). Then we should also have f^{n+1}, g^{n+1}, h^{n+1} and a new set of solutions if and only if the overdetermined set of equations obtained by the WTC analysis is satisfied. So we now use $\left(f^n, g^n, h^n\right)$ instead of (f, g, h) in (3.5.17) and use equation (3.5.19) to obtain

$$D_t^2\left(\phi_{n-1}f^{n-1}\cdot\phi_{n-1}f^{n-1}\right) = -D_x^2\left(\phi_{n-1}f^{n-1}\cdot\phi_{n-1}f^{n-1}\right)$$

$$-1/3D_x^4\left(\phi_{n-1}f^{n-1}\cdot\phi_{n-1}f^{n-1}\right)+\left(\phi_{n-1}f^{n-1}\cdot\phi_{n-1}f^{n-1}\right)$$

$$=\left(\psi_0 f^{n-1}+\phi_{n-1}g^{n-1}\right)\left(\chi_0 f^{n-1}+\phi_{n-1}h^{n-1}\right)$$

or

$$\left\{2\left[\phi_{(n-1)}\phi_{(n-1)_{tt}}-\phi_{(n-1)t}^2\right]\left(f^{n-1}\right)^2+\phi_{n-1}^2 D_t^2\left(f^{n-1}f^{n-1}\right)\right\}$$

$$-\left\{2\left[\phi_{n-1}\phi_{(n-1)xx}-\phi_{(n-1)x}^2\right]\left(f^{n-1}\right)^2+\phi_{n-1}^2 D_x^2\left[f^{n-1}f^{n-1}\right]\right\}$$

$$-1/3\left[2\phi_{n-1}\phi_{(n-1)xxxx}-8\phi_{(n-1)x}-\phi_{(n-1)xxx}+6\phi_{(n-1)xx}^2\right]\left(f^{n-1}\right)^2$$

$$-2\left[2\phi_{n-1}\phi_{n-1xx}-2\phi_{(n-1)x}^2\right]\left[2f^{n-1}f_{xx}^{n-1}-2\left(f_x^{n-1}\right)^2\right]$$

$$-1/3\phi_{n-1}^2 D_x^4\left(f^{n-1}f^{n-1}\right)+\left(\phi_{n-1}f^{n-1}\cdot\phi_{n-1}f^{n-1}\right)$$

$$=\psi_0\chi_0\left(f^{n-1}\right)^2+\left(f^{n-1}\right)^2\phi_{n-1}\left[\chi_0\left(g^{n-1}/f^{n-1}\right)+\psi_0\left(h^{n-1}/f^{n-1}\right)\right]$$

$$+\phi_{n-1}^2 g^{n-1}h^{n-1}$$

or

$$2\left[\phi_{n-1}\phi_{(n-1)tt}-\phi_{(n-1)t}^2\right]\left(f^{n-1}\right)^2-2\left[\phi_{n-1}\phi_{(n-1)xx}-\phi_{(n-1)x}^2\right]\left(f^{n-1}\right)^2$$

$$-1/3\left[2\phi_{(n-1)}\phi_{(n-1)xxxx}-8\phi_{(n-1)x}\phi_{(n-1)xxx}+6\phi_{(n-1)xx}^2\right]\left(f^{n-1}\right)^2$$

$$-2\left[2\phi_{n-1}\phi_{(n-1)xx}-2\phi_{(n-1)x}^2\right]\left[2f^{n-1}f_{xx}^{n-1}-2\left(f_x^{n-1}\right)^2\right]$$

$$+\left[\phi_{n-1}^2\left(D_t^2-D_x^2-1/3D_x^4+\rho\right)\times\left(f^{n-1}f^{n-1}\right)-\phi_{n-1}^2 g^{n-1}h^{n-1}\right]$$

$$=\psi_0\chi_0\left(f^{n-1}\right)^2+\left(f^{n-1}\right)^2\phi_{n-1}\left[\chi_0\left(g^{n-1}/f^{n-1}\right)+\psi_0\left(h^{n-1}/f^{n-1}\right)\right].$$

Now the terms in the last bracket on the left hand side is 0, and if we divide by $\left(nf^{n-1}\right)^2$ on both sides, we get

$$2\Big[\phi_{(n-1)}\phi_{(n-1)tt} - \phi_{(n-1)t}^2\Big] - 2\Big[\phi_{(n-1)}\phi_{(n-1)xx} - \phi_{(n-1)x}^2\Big]$$

$$-\frac{1}{3}\Big[2\phi_{(n-1)}\phi_{(n-1)xxxx} - 8\phi_{(n-1)x}\phi_{(n-1)xx} + 6\phi_{(n-1)xx}^2\Big]$$

$$-2\Big[2\phi_{(n-1)}\cdot\phi_{(n-1)xx} - 2\phi_{(n-1)x}^2\Big]\cdot\Big[2f^{n-1}f_{xx}^{n-1} - 2\big(f_x^{n-1}\big)^2\Big]\Big[1\Big/\big(f^{n-1}\big)^2\Big]$$

$$= \psi_0\chi_0 + \phi_{n-1}\cdot\Big[\chi_0\big(g^{n-1}/f^{n-1}\big) + \psi_0\big(h^{n-1}/f^{n-1}\big)\Big] .$$

Now it is easy to observe

$$\frac{2\Big[f^{n-1}f_{xx}^{n-1} - \big(f_x^{n-1}\big)^2\Big]}{\big(f^{n-1}\big)^2} = 2\big(\ln f^{n-1}\big)_{xx} = u_2 .$$

And then writing ϕ for ϕ_{n-1}, ψ_1 for f^{n-1}/f^{n-1} and χ_1 for h^{n-1}/f^{n-1} , we have

$$2\big(\phi\phi_{tt} - \phi_t^2\big) - 2\big(\phi\phi_x - \phi_x^2\big) - 1/3\big(2\phi\phi_{xxxx} - 8\phi_x\phi_{xxx} + 6\phi_{xx}^2\big)$$

$$-2\big[2\phi\phi_{xx} - 2\phi_x^2\big]u_2 = \phi(\chi_0\psi_1 + \psi_0\chi_1) + \psi_0\chi_0 .$$

So we can write

$$6\big(\phi\phi_{tt} - \phi_t^2\big) - 6\big(\phi\phi_x - \phi_x^2\big) - \big(2\phi\phi_{xxxx} - 8\phi_x\phi_{xxx} + 6\phi_{xx}^2\big)$$

$$-6\big[2\phi\phi_{xx} - 2\phi_x^2\big]u_2 = 3\psi_0\chi_0 + 3\phi(\chi_0\psi_1 + \psi_0\chi_1)$$

so that

$$\phi\big[6\phi_{tt} - 6\phi_{xx} - 2\phi_{xxxx} - 12\phi_{xx}u_2 - 3(\chi_0\psi_1 + \psi_0\chi_1)\big]$$

$$-\big[6\phi_t^2 - 6\phi_x^2 - 8\phi_x\phi_{xxx} + 6\phi_{xx}^2 - 12u_2\phi_x^2 - 12\phi_x^2u_2 + 3\psi_0\chi_0\big] = 0 . \qquad (3.5.19b)$$

This equation is identically satisfied if each of the bracketed expressions vanishes, which is the case due to the overdetermined set obtained above. Let us now differentiate (3.5.8) and multiply by ϕ_x to get

$$12\phi_x^3 u_{2x} = -24\phi_x^2 \phi_{xx} u_2 + 12\phi_x \phi_t \phi_{xt} - 12\phi_x^2 \phi_{xx}$$
$$+ 4\phi_x \phi_{xx} \phi_{xxx} - 8\phi_x^2 \phi_{xxxx} + 3\phi_x (\psi_0 \chi_0)_x \; . \tag{3.5.20}$$

And multiplying the same by ϕ_{xx} we get

$$6\phi_{xx}\phi_t^2 = 12\phi_x^2 \phi_{xx} u_2 + 6\phi_x^2 \phi_{xx} + 8\phi_x \phi_{xx} \phi_{xxx} - 6\phi_{xx}^3 - 3\phi_{xx} \cdot (\psi_0 \chi_0) \tag{3.5.21}$$

whence using (3.5.20), (3.5.21) and (3.5.9) we arrive at

$$6\phi_{tt} - 6\phi_{xx} - 2\phi_{xxxx} - 12\phi_{xx} u_2 - 3(\chi_0 \psi_1 + \psi_0 \chi_1) = 0 \; .$$

It is then easy to observe that equation (3.5.19a) is identically satisfied.

Soliton solution
Since we have already observed that we can throw our equation in Hirota's form, it will be very easy to construct the soliton solution if we substitute the following expansion in equations (3.5.17) and (3.5.18)

$$f^{(0)} = 1 + \varepsilon f_1 + \varepsilon^2 f_2 + \dots$$
$$g^{(0)} = A e^{i(kx - \omega t)} \left[1 + \varepsilon g_1 + \varepsilon^2 g_2 + \dots \right]$$
$$h^{(0)} = B e^{-i(kx - \omega t)} \left[1 + \varepsilon h_1 + \varepsilon^i h_1 + \dots \right]$$

and compare various powers of ε. A simple computation immediately leads to

$$f_1 = -h_1 = -g_1$$
$$f_{1tt} - f_{1xx} - 1/3 f_{1xxxx} + 2\rho f_1 = 0 \; .$$
$$f_{1x}^2 - f_1 f_{1xx} = 0$$

A general solution is written as

$$f_1 = p(x) q(t) \cdot p e^{ax - bt} + Q e^{ax - bt} \; ,$$

where $b = (k - 2\rho)^{1/2}$

$$a^2 + 1/3sa^4 = K \ ,$$

from which one obtains at once

$$
\begin{aligned}
u &= \frac{2a^2 p e^{ax+bt}}{\left(1 + p e^{ax+bt}\right)^2} \\[2mm]
&= \frac{A\left[1 - p e^{ax+bt}\right]e^{i(kx-\omega t)}}{\left(1 + p e^{ax+bt}\right)} \\[2mm]
&= \frac{B\left[1 - p e^{ax+bt}\right]e^{-i(kx-\omega t)}}{\left(1 + p e^{ax+bt}\right)}
\end{aligned}
$$

with

$$\omega - k^2 + \lambda = 0$$

which is nothing but the one soliton solution.

Our above discussion brings out an important feature of the WTC approach to Painleve analysis. The method is useful even for a system for which the usual Lax pair is not known to exist. In the present situation, one branch has passed all the criteria of the test, and it is proved that the system is completely integrable. The Hirota method, when coupled with the WTC approach, helps to prove the recursive nature of the BT by induction. Lastly, the soliton solutions can be constructed explicitly. Of course not all of the equations whose Lax pairs are unknown can have such regular properties. The following provides examples of two equations that are not completely integrable, but the Painleve test can help to construct some special type of solution of such systems.

The Newell-Whitehead equation

Another interesting equation is the Newell-Whitehead equation written as

$$u_t = u_{xx} + u - 2u^3 \ . \tag{3.5.22}$$

A WTC expansion is obtained by setting

$$u = \phi^{-1} \sum_{j=0}^{\infty} u_j \phi^j \ . \tag{3.5.23}$$

The leading resonance is $j = 4$. The recursion relations determine the coefficients, which can be conveniently written in terms of

$$\omega = \phi_t / \phi_x \quad \text{and} \quad v = \phi_{xx} / \phi_x .$$

For example,

$j = 0 , \quad u_0 = \phi_x$

$j = 1 , \quad u_1 = 1/6(\omega - 3v)$

$j = 2 , \quad u_2 = 1/6\phi_x\left(v_x - \omega_x + 1 - 1/6\omega^2 - 1/2v^2\right)$

$j = 3 , \quad u_3 = 1/24\phi_x^2\{-\omega_t + 2\omega_{xx} + 7/3\omega\omega_x - 2\omega$
$\qquad\qquad\qquad +4/9\omega^3 - v_{xx} + 3vv_x + 2v - v^3 - 2v\omega_x - 1/3\omega^2v\}$

$j = 4 , \quad (j+1)(j-4)u_4 = 1/3\phi_x^2\left(\omega_t - \omega\omega_x + 3/2\omega - 1/3\omega^3\right) .$ (3.5.24)

Because the resonance at $j = 4$ in the left side of the fifth equation in (3.5.24) is 0, we must have

$$\omega_t - \omega\omega_x + 3/2\omega - 1/3\omega^3 = 0 .$$ (3.5.25)

So that we may say that the equation has conditional Painleve property. In other words, this implies that the variable u can have an expansion of the form (3.5.23) if and only if ϕ satisfies a condition of the form (3.5.25), written in terms of the new variable ω. Just like the Burger equation, equation (3.5.2) does possess shock like solutions. There exists two constant solutions $\omega = 0$, $\pm 3/2$. Because $\omega = \phi_t / \phi_x$, it is easy to observe that $\omega = 0$ corresponds to $\phi = \phi(x)$, and we can now drop the u_t term in (3.5.25) so that

$$u_{xx} + u - 2u^3 = 0$$

soluble in terms of Jacobi elliptic functions. The constant solution $\omega = \pm 3/2$ correspond to $\phi(x \pm 3/2t)$ so that we get an expansion of u in the variable $x \pm (3/2)t$. So if we set $z = x \pm \left(\dfrac{3}{2}\right)t$ we get

$$u_{zz} \mp 3/2u_z + u - 2u^3 = 0 .$$

Set now $u = \exp(\pm z/\sqrt{2})\psi\left(\pm\sqrt{2}\exp(\pm 2/v_2)\right)$. This is transformed to

$$\psi'' = 2\psi^3 \, ,$$

which is integrable also in terms of Jacobi elliptic functions.

If one now attempts to truncate the expansion (3.5.23)

$$u = \pm \phi_x/\phi + u_1 \, , \tag{3.5.25a}$$

then following the usual procedure

$$\phi_t - 3\phi_{xx} \mp 6\phi_x u_1 = 0 \tag{3.5.26a}$$

$$-\phi_{xt} + \phi_{3x} + \phi_x - 6\phi_x u_1^2 = 0 \tag{3.5.26b}$$

$$-u_{1t} + u_{1xx} + u_1 - 2u_1^3 = 0 \tag{3.5.26c}$$

gives an overdetermined set for ϕ. In terms of variable ω and v, equation (3.5.26a) can be written as

$$u_1 = \pm\left(\frac{1}{6} - \frac{1}{2}v\right) \, .$$

Putting this expression for u_1 in (3.5.26b) and (3.5.26c), we get

$$\omega_x + 1/6\omega^2 - 1 = v_x - 1/2v^2 \tag{3.5.27}$$

and

$$\frac{1}{3}\omega_t - \frac{1}{3}\omega_{xx} + \frac{1}{3}\omega + \frac{1}{54}\omega^3 - v_t + v_{xx} + v - \frac{1}{2}v^3 - \frac{1}{6}\omega^2 v + \frac{1}{2}\omega v^3 = 0 \, . \tag{3.5.28}$$

In addition, there is the natural relationship between v and ω,

$$v_t = v_{xx} + \omega v_x + \omega_x v \, . \tag{3.5.29}$$

If we take the time derivative of (3.5.28) and combine it with (3.5.29), we get

$$-\omega_{xt} - \frac{1}{3}\omega\omega_t + \omega_{xxx} + \omega\omega_{xx} + \frac{2}{3}\omega^2\omega_x + 2\omega_x^2 - 2\omega_x = 0 . \tag{3.5.30}$$

The terms containing v can be eliminated from (3.5.30) to yield

$$-\omega_t + \omega_{xx} - 2\omega + \frac{4}{9}\omega^3 + 2\omega\omega_x = 0 . \tag{3.5.31}$$

If equation (3.5.31) is subtracted from the x derivative of (3.5.28), a Burger equation results,

$$\omega_t + 2\omega\omega_x + 3\omega_{xx} = 0 , \tag{3.5.32}$$

which reduces (3.5.31) to

$$\omega_{xx} + \omega\omega_x - \frac{1}{2}\omega + \frac{1}{9}\omega^3 = 0 .$$

Set $\omega = 3f_x/f$ so that

$$f_{xxx} - \frac{1}{2}f_x = 0 .$$

Solution of this linear equation is straightforward and leads to

$$\omega = \frac{3}{\sqrt{2}} \frac{A_0 e^{x/\sqrt{2}} - B_0 e^{-x/\sqrt{2}}}{A_0 e^{x/\sqrt{2}} + B_0 e^{-x/\sqrt{2}} + K_0} .$$

A_0, B_0, K_0 are functions of time. This time dependence is governed by the equation (3.5.32) and it is found that

$$\omega = 3/2 \frac{A\exp(x/\sqrt{2}) + B\exp(-x/\sqrt{2})}{A\exp(x/\sqrt{2}) + B\exp(-x/\sqrt{2}) + K\exp(3t/2)} .$$

Now we can find out the form of ϕ from the relation $\omega = \phi_t/\phi_x$. It is observed that $\phi = \phi(F)$.

$$F = \frac{-2\,2\exp\left(-\frac{3}{2}t\right) - \sqrt{2}K/\,A\exp(-x/\sqrt{2})}{A\exp(x/\sqrt{2}) - B\exp(-x/\sqrt{2})}$$

But equation (3.5.27) still has to be satisfied. It can be rewritten as

$$\{\phi, x\} = \omega_x + \frac{1}{6}\omega^2 + 1 \ .$$

Note that $\{\phi, x\}$ stands for the Schwarzian derivative. A property of this Schwarzian is that

$$\{\phi, x\} = \{\phi, F\}F_x^2 + \{F, x\} \ , \tag{3.5.32a}$$

and the above equation (3.5.32a) can be transformed to

$$\{\phi_i, F\}F_x^2 = \left(\omega_x + \frac{1}{6}\omega^2 - 1\right) - \{F, x\} \ .$$

Using the functional form of $\{x, F\}$ and F, the right hand side of this is seen to be 0, so we are left with

$$\{\phi, F\} = 0 \ ,$$

which integrates to

$$\phi = \frac{C_1}{F + C_2} + C_3 \ .$$

C_1, C_2, C_3 are constants. Solution u can be explicitly obtained, so here we have an example that is second order but possesses only one resonance. It is not completely integrable, but some special solutions can be constructed through WTC expansion. Though equation (3.5.25a) looks like a Backlund transformation, it is found that its use does not lead to any new solution. For example, if we start from $u_1 = 0$, then

$$u = \phi_x/\phi \ ,$$

and we get

$$\phi_t - 3\phi_{xx} = 0 , \qquad -\phi_{xt} + \phi_{xxx} + \phi_x = 0 ,$$

which can be easily solved for ϕ. And we find the same functional form as obtained from equation (3.5.23). Lastly, it may be remarked that equation (3.5.25a) with $u_1 = 0$ is exactly Hirota's substitutions used by Kawahara and Tanaka.

We can now try to observe what happens if we want to retain all the terms in WTC expansion, namely

$$u = \phi_x/\phi + \sum_{j=4}^{\infty} u_j \phi^j .$$

Such a series is still very complicated because of the complicated dependence of u_5, u_6, etc. on various derivatives of u_4. Alternatively,

$$u = \phi_x/\phi + u_4 \phi^3 + u_8 \phi^7 + u_{12} \phi^{11} + \dots .$$

Such an expansion will be self-consistent if

$$u_{4k} = C_{4k} \phi_x .$$

To prove it, we write the recursion relation obtained from (3.5.22)

$$\left[(j-4)(j+1)\phi_x^2 \right] u_j = -u_{j-2,t} - (j-2)u_{j-1}\phi_t + u_{j-2,xx} + 2(j-2)u_{j-1,x}$$

$$+ (j-2)u_{j-1}\phi_{xx} + u_{j-2} - \sum_{k=1}^{j-1} \sum_{l=0}^{k} u_{j-k} u_{k-1} u_1 .$$

For the truncated case, set $u_1 = u_2 = u_3 = 0$. Whence we get conditions

$$\phi_t = 3\phi_{xx} \tag{3.5.33a}$$

$$\phi_{tx} = \phi_{xxx} + \phi_x . \tag{3.5.33b}$$

The first non-zero coefficient is u_4. The next, u_5, will also be non-zero unless

$$-u_4\phi_t + u_{4x}\phi_x + u_4\phi_{xx} = 0 \; . \tag{3.5.34}$$

Nonlinearity ensures that each multiple of the resonance that is u_{4k} is non-zero and each following coefficient u_{4k+1} is also non-zero,

$$-u_{4k}\phi_t + 2u_{4k,x}\phi_x + u_{4k}\phi_{xx} = 0$$

on the assumption that $u_{4k-1} = 0$. From these equations it is easy to demonstrate that

$$u_{4k} = C_{4k}\phi_x \; .$$

This condition implies that

$$u = \phi_x/\phi\left(1 + C_4\phi^4 + C_8\phi^8 + C_{12}\phi^{12} + ...\right)$$

$$= \phi_x/\phi f(\phi) \; , \tag{3.5.35}$$

where $f(\phi)$ represents any analytic function of ϕ. Regarding the factor ϕ_x/ϕ as the singular part of u is highly reminiscent of the rescaling ansatz used to analyse the logarithmic psi-series that arises in nonintegrable systems. Now equations (3.5.33a) and (3.5.33b) lead to

$$\phi = A\exp((x + 3t/\sqrt{2})/\sqrt{2}) + B\exp(-(x - 3t/\sqrt{2})/\sqrt{2}) + C \; ,$$

where A, B and C are constants. With this information, if the ansatz (3.5.35) are substituted in (3.5.34), one gets

$$f\left\{\phi_{xxx}/\phi - 3\phi_x\phi_{xx}/\phi^2 + 2\phi_x^3/\phi^3 + \phi_x/\phi - \phi_{xt}/\phi + \phi_x\phi_t/\phi^2\right\}$$
$$+ f'\left\{3\phi_x\phi_{xx}/\phi - 2\phi_x^3/\phi^2 + \phi_x\phi_t/\phi\right\} + f''\left\{2\phi_x^3/\phi^3\right\} - 2f'^3\left\{2\phi_x^3/\phi^3\right\} = 0 \; .$$

Using again the equations satisfied by ϕ, this complicated equation reduces to

$$\phi^2 f - 2\phi f + 2f + 2f^3 = 0 \; ,$$

which turns into

$$F + F^3 = 0$$

if we use $F = f(\phi)/\phi$.

One should note that earlier we obtained an equation similar to this one for ψ. This equation is again solvable by Jacobi elliptic function, so we have

$$u = \phi_x F(\phi)$$
$$= 2[A\exp(x + 3t/\sqrt{2})/\sqrt{2} - B\exp\{-(x - 3t/\sqrt{2})/\sqrt{2}\}]$$
$$\times ds(\sqrt{2}(A\exp[\{x + 3t/\sqrt{2}\}/\sqrt{2}] + B\exp[-\{x - 3t/\sqrt{2}\}/\sqrt{2}]$$

ds is the elliptic function with modulas $1/2$.

The Ginsburg-Landou equation[72]

One of the most important equations in physics is the so-called Ginsburg-Landou system, which is used in elementary particle physics, phase transition, and in explaining various other phenomena. There are two variations of it: one with the mass term and the other without it. Here we consider the situation with a mass-like term. The equation under consideration is

$$iu + pu_{xx} + q|u|^2 u = ir\,u \,, \tag{3.5.36}$$

where $p = p_r + ip_i$, $q = q_r + iq_i$, r is real.

We can also consider the following two coupled set:

$$iu_t + pu_{xx} + qu^2 v = iru$$
$$-iv_t + \tilde{p}v_{xx} + \tilde{q}v^2 u = -irv \,.$$

$\tilde{p}, \tilde{q}$ are complex conjugate of p, q. In the leading order, we set

$$u \sim u_0 \phi^\varepsilon \,, \qquad v \sim v_0 \phi^\eta \,.$$

Equating the exponents of the dominant order we find

$$\varepsilon + \eta = -2 \ .$$

(3.5.37)

The corresponding coefficients lead to

$$u_0 v_0 = - \frac{p}{q} \varepsilon(\varepsilon - 1)\phi_x^2$$

and

$$u_0 v_0 = - \frac{\tilde{p}}{\tilde{q}} \eta(n-1)\phi_x^2 \ ,$$

so we get at once

$$p/q\varepsilon(\varepsilon - 1) = p/q\eta(\eta - 1) \ .$$

Solution for ε and n gives

$$\varepsilon = -1 - i\alpha, \qquad n = -1 + i\alpha \ ,$$

where α is a root of

$$\alpha^2 - 3\alpha\beta - 2 = 0$$

$$\beta = \frac{\mathrm{Re}(p/q)}{\mathrm{Im}(p/q)} \ .$$

The expansion over the singular manifold takes the form

$$u = \phi^{-i\alpha}\left(u_0/\phi + u_1 + ...\right)$$
$$v = \phi^{+i\alpha}\left(v_0/\phi + v_1 + ...\right) \ .$$

In the next stage, we proceed to search for the resonances from the recursion conditions given below

$$T\begin{pmatrix} u_j \\ v_j \end{pmatrix} = \begin{pmatrix} G_j \\ H_j \end{pmatrix} ,$$

with

$$T = \begin{pmatrix} p(j-1-i\alpha)(j-2-i\alpha)\phi_x^2 + 2qu_0v_0 & qu_0^2 \\ \tilde{q}v_0^2 & \tilde{p}(j-1+i\alpha)(j-2+i\alpha)\phi_x^2 + 2\tilde{q}u_0v_0 \end{pmatrix}$$

$\det(T) = 0$ leads to the resonance positions

$$j = -1, \quad 0, \quad 7/2 \pm 1/2\sqrt{1 - 24\alpha^2} .$$

When $\alpha = 0$, these are all integers $(-1, 0, 3, 4)$. The truncated expansion can also be written as

$$u = u_0 / \phi^{1+i\alpha} + u_1 / \phi^{i\alpha} ; \quad u = v_0 / \phi^{1-i\alpha} + v_1 / \phi^{-i\alpha} . \tag{3.5.38}$$

Note that they are usually truncated at the constant level term, but that is not the case here. Also, equation (3.5.38) is not so useful as in the previous cases. But if we note that u_1, v_1 again satisfies the same set of equations (3.5.36), we can start with $u_1 = v_1 = 0$ and substitute (3.5.38) in equation (3.5.36). The following equations are obtained at each power of ϕ (here $u_0 = a$, $v_0 = b$):

$$\phi^{-3} \pm i\alpha : p(1+i\alpha)(2+i\alpha)\phi_x^2 + qab = 0$$
$$\tilde{p}(1-i\alpha)(2-i\alpha)\phi_x^2 + \tilde{q}ab = 0$$

$$\phi^{-2} \pm i\alpha : ia\phi_t + pa\phi_{xx} + 2pa_x\phi_x = 0 \tag{3.5.39a}$$
$$-ib\phi_t + \tilde{p}b\phi_{xx} + 2\tilde{p}b_x\phi_x = 0 \tag{3.5.39b}$$

$$\phi^{-1} \pm i\alpha : ia_t + pa_{xx} = i\gamma a \tag{3.5.39c}$$
$$-ib_t + \tilde{p}b_{xx} = -i\gamma b . \tag{3.5.39d}$$

This set of equations is an over-determined set for a, b, and ϕ. Due to the relation (3.5.37) between ε and n, one can at once deduce

$$ab = K\phi_x^2 , \tag{3.5.39}$$

where K is a real constant. An important consequence of this equation is

$$a_x/a + b_x/b = 2\phi_{xx}/\phi_x . \qquad (3.5.40)$$

Also, from equations (3.5.38a,b), one gets

$$p_i\phi_t + |p|^2 \phi_x(a_x/a + b_x/b) + |p|^2 b_{xx} = 0 ,$$

which, when considered along with (3.5.40), gives a linear equation for ϕ:

$$p_i\phi_t + 3|p|^2 \phi_{xx} = 0 .$$

We now use this equation to eliminate ϕ_t from (3.5.38a,b) so that

$$\frac{a_x}{a} = (1+i\beta)\frac{\phi^\phi xx}{\phi_x} ; \qquad \frac{b_x}{b} = (1-i\beta)\frac{\phi_{xx}}{\phi_x} ,$$

where $\beta = 3p_r/2p_i$. These can be integrated at once, and we get

$$a = f(t)\phi_x^{1+i\beta} ; \qquad b = g(t)\phi_x^{1-i\beta} ,$$

where f and g are functions to be determined. Note that equation (3.5.39) demands

$$fg = K ,$$

whence

$$f_t/f + g_t/g = 0 .$$

There are other conditions due to equations (3.5.39c,d) that ϕ must satisfy. If the forms of a and b are substituted in equations (3.5.39c) and (3.5.39d), then

$$\frac{i^f t}{f} + (1+i\beta)\left(p - 3i|p|^2/p_i\right)\frac{\phi_{xxx}}{\phi_x} + p(i\beta)(1+i\beta)(\phi_{xx}/\phi_x)^2 = ir$$

$$\frac{i^f t}{f} + (1-i\beta)\left(\tilde{p} + 3i|p|^2/p_i\right)\frac{\phi_{xxx}}{\phi_x} - \tilde{p}(i\beta)(1-i\beta)(\phi_{xx}/\phi_x)^2 = -ir .$$

The common solution can be written as

$$\phi = C_1(t)\exp(Gx) + C_2(t)$$

$$G^2 = -4rp_i \Big/ \left(8p_i^2 + qp_r^2\right) ,$$

so that

$$f_t/f = -ir\frac{10p_i^2 p_r + qp_r^3}{8p_i^3 + qp_r^2 p_i} = i\omega .$$

At the final step we require ϕ to satisfy the linear equation, thus determining C_1 and C_2. So finally

$$\phi = C_3\exp\left(G^2 Dt + Gx\right) + C_4 ,$$

where $D = -3|p|^2/p_i$ and C_3, C_4 are constants. Collecting all the information the solution 4 can be presented as

$$u = g\exp[i(kx-t)]\frac{\exp[-k(x-ct)]}{\{1+\exp(-k(x-Ct))\}}1+i\alpha ,$$

where

$$k = G , \quad C = GD$$

$$\Omega = iG^2 D(1+i\beta) - \omega$$

$$k = G(1+i\beta) ; \quad |g| = G\sqrt{K} .$$

In the above discussions we have essentially followed the two excellent articles by Cariello and Tabor.[72] These two equations exhibit some novel features that are not usually displayed by many. They also serve to show the utility of the Painleve expansion, truncation even, in case of nonintegrable systems. The occurrence of complex exponent is also a new feature. In the following we continue our analysis with respect to one more system — the Kuramoto-Shivashinsky (KS) equation, which has played an important role in various nonlinear phenomena. It is interesting to observe that though KS equation is known to exhibit chaos, the Painleve test can extract lots of information from it.

Kuramoto-Shivashinsky equation[58,59,60]

The KS equation is encountered in the study of continuous media. It is written as

$$u_t + uu_x + \mu u_{xx} + vu_{xxxx} = 0 . \tag{3.5.41}$$

It describes the bifurcation structure in complex fluid flow, the fluctuation of the position of a flame, the motion of a fluid dripping down a vertical wall, and a uniform oscillating chemical reaction in a homogeneous medium. As per the WTC prescription, one sets

$$u(x,t) = \phi(x,t)^p \sum_{j=0}^{\infty} u_j(x,t) .$$

$p < 0$, $\phi(x,t) = 0$ is the equation of a noncharacteristic singular manifold, the function u_j is determined by substituting this series in equation (3.5.41). Recall that we previously wrote the result of such a substitution as

$$\sum_{j=0}^{\infty} F_j\left(u_0,...,u_j, \phi\right)\phi^{j+q} = 0 ,$$

where F_j depends on ϕ and its various derivatives, and q is a negative integer. In the present situation, the leading power turns out to be $p = -3$ and

$$u_0 = 120\phi_x^3 v ,$$

with

$$P(j) = (j-6)\left(j^2 - 13j + 60\right) ,$$

so that the resonances occur at $j = 6$ and $j = \dfrac{13}{2} \pm i\dfrac{\sqrt{71}}{2}$. The situation is similar to the case of capillary gravity waves, where we also lost two resonances. So in principle, there is no way to obtain the most general solution. But to study any special solution, we compute the coefficients up to $j = 6$ and get

$$u_1 = -180v\phi_x\phi_{xx}$$

$$u_2 = 60\phi_{xxx}v + \frac{60}{19}\mu\phi_x$$

$$u_3 = -\phi_t/\phi_x + 15v\left\{-\phi_{xxxx}/\phi_x + 2\cdot\phi_{xx}\phi_{xxx}/\phi_x^2 - \phi_{xx}^3/\phi_x^3\right\} - 30/19\mu\cdot\phi_{xx}/\phi_x .$$

It can be proved that the resonance at $j = 6$ is compatible. The corresponding compatibility condition has the form

$$F_6 = -\left(\phi_x^{-1}E_5\right)_x - 1/2\left(\phi_x^{-1}\left(\phi_x^{-1}E_4\right)_x\right)_x - 1/6\left(\phi_x^{-1}\left(\phi_x^{-1}\left(\phi_x^{-1}E_3\right)_x\right)_x\right)_x ,$$

where u_0, u_1, u_2 have been replaced by their values. So, we can at best obtain a solution depending upon two arbitrary functions ϕ and u_6. Next we truncate the series at $j = 3$,

$$u = 60v(\log \phi)_{xxx} + 60/19\mu(\log \phi)_x + u_3 ,$$

provided u_3 and ϕ satisfy

$$F_j(\phi, u_3) = 0 \quad j = 3 \text{ to } 7 .$$

Due to the built-in conformal invariance in the Painleve expressions, we can also work with $\phi_x^{j-7}F_j$ instead of F_j. Equation $F_3 = 0$ gives the expression for u_3. As noted above, F_6 involves F_3, F_4, and F_5. Equation $F_7 = 0$ is just the KS equation for u_3. Substituting the expression of u_3 in all of these yields four equations depending on ϕ and ϕ_x. Note that F_4 is invariant under the homographic transformation, and the same is true for the newly defined $\tilde{F}_j$

$$\tilde{F}_j = \sum_{k=4}^{j}(\phi_{xx}/2\phi_x)^{j-k}\frac{(7-K)!}{(7-j)!(j-k)!}F_K ; \quad j = 4 \text{ to } 7 .$$

As explained in Section 3.2, it is because of this fact that all the expressions F_j can be expressed in terms of S and C. These are

$$u_3 = C - 15vS_x - 30v(S + \mu/19v)\phi_{xx}/\phi_x - 15v(\phi_{xx}/\phi_x)^3$$

$$F_4 = 120\left(6v^2S_{xx} + 4v^2S^2 + \frac{20\mu v}{19}S - 2vC_x - 11\mu^2/19^2\right) = 0$$

$$F_5 = -60\left(7v^2S_{xx} + 3v^2S^2 + 25\mu v/19S - 3vC_x\right)_x = 0$$

$$F_7 = C_t + CC_x + \frac{177\mu}{38}C_{xx} + f\left(\tilde{F}_4, \tilde{F}_5\right) + v\left(45SC_{xx} + 12C_{xxxx} + 30C_xS_x\right) = 0 ,$$

where f is an expression $\rightarrow 0$ as F_4, and $F_5 \rightarrow 0$.

Integrating $\tilde{F}_5 = 0$ we get

$$7v^2 S_{xx} + 3v^2 S^2 + \frac{25\mu v}{19} S - 3vC_x - \Lambda = 0$$

Λ is a function of t only. Eliminating C_x between $F_4 = 0$ and $F_5 = 0$ we get

$$4v^2 S_{xx} + 6v^2 S^2 + \frac{10\mu v}{19} S + 2\Lambda - \frac{33\mu^2}{192} = 0 .$$

This is a second-order ordinary differential equation for S, and it has two types of solutions

$$S = - \frac{5\mu}{114v} - 4\wp ,$$

where $\wp$ is the Weierstrass elliptic function.

$\wp(x - x_0(t), \, g_2(t), \, g_3(t))$, x_0, g_3 are arbitrary functions and

$$g_2 = - \frac{\Lambda}{4v^2} + \frac{223\mu^2}{48(19)^2 v^2} .$$

The associated value of C_x is

$$C_x = -40v\wp^2 - 80\mu/57\wp + 6vg_2 - 112\mu^2/9.19^2 v$$

which on integration yields

$$C = - \frac{20v}{33}\wp' + \frac{80\mu}{57}\zeta - \left\{ \frac{112\mu^2}{9.19^2 v} - \frac{8}{3}vg_2 \right\}(x - x_0) + g_4 ,$$

in which ζ is another Weierstrass elliptic function. $\wp'$ is the derivative of $\wp$ with respect to its first argument.

The other is the singular solution $S_{xx} = 0$, whence S is one of the roots of

$$S^2 + \frac{5\mu}{3.19v} S + \frac{\Lambda}{3v^2} - \frac{11\mu^2}{2.19^2 v^2} = 0 \ ,$$

so that S is independent of x. Using this,

$$C = 2vx\left\{ S^2 + \frac{5\mu}{19v} S - \frac{11\mu^2}{4.19^2 v^2} \right\} + K_1$$

$$S'(t) = -4vS\left\{ S^2 + \frac{5\mu}{19v} S \frac{11\mu^2}{4.19^2 v^2} \right\}$$

$$\Lambda = -3v^2 \left\{ S^2 + \frac{5\mu}{57v} S - \frac{11\mu^2}{2.19^2 v^2} \right\} .$$

Since function C is linear in x and S is independent of x, equation $\tilde{F}_7 = 0$ reduces to

$$C_t + CC_x = 0$$

and implies that

$$S^2 + \frac{5\mu}{19v} S - \frac{11\mu^2}{4.19^2 v^2} = 0 \ .$$

The two values of S are

$$S_1 = 1/38 \cdot \mu / v$$

and

$$S_2 = -11/38 \cdot \mu / v \ .$$

From the general theory of nonlinear differential equation, it is known that the solution of the third order nonlinear equation $\{\phi, x\} = r$ is

$$\phi = \frac{\alpha y_1 + \beta y_2}{\gamma y_1 + y_2}$$

α, β, γ, arbitrary constants, $\alpha - \beta\gamma \neq 0$. y_1, y_2 are two independent solutions of

$$y'' + \frac{r}{2}y = 0 .$$

Since $K^2 = -2S$, we get that

$$\phi = \frac{\alpha + \beta\exp(K(x-ct))}{\gamma + \delta\exp(K(x-ct))} .$$

It will be very pertinent to point out that though we have obtained a Painleve-Backlund-type transformation, but it is not possible to iterate this relation by choosing u_3 and ϕ independently, and the class of solution remains very much restricted.

Conclusion

In the above discussions we have presented five examples of nonintegrable nonlinear PDEs in the light of the Painleve analysis. Each presents a different situation, so the analysis varies from one to the other. The most important idea that emerges out of this analysis is that even in such situations it will be possible to derive many results via the WTC-type expansion. But one should be careful and keep in mind that the consequence of such analysis may be different in different problems.

3.6 Painleve analysis in more than $(1+1)$ dimension

After the initial success of the various techniques in case of nonlinear systems in $(1+1)$ dimension, there was vigorous efforts to apply such methodologies in more than the $(1+1)$ dimensional situation. The methods of Lie symmetries, prolongation structure, Bäcklund transformations, etc. showed good promise for the higher dimensional cases, so it was felt that the proper extension of the Painleve test should be possible. Later it was found that certain integrable systems in $(2+1)$ dimension – such as KP equation, NLS (nonlinear Schrödinger) in $(2+1)$ dimension, and Burger equation in $(3+1)$ – can be analyzed by the singular manifold approach, and many properties can be studied even in these higher dimensional situations. In the following, we shall consider the cases of NLS in $(2+1)$ dimension, Melnikov's equation, Burger in $(3+1)$ dimension, and that of a nonrelativistic field theory in $(2+1)$ dimension. The first three cases are completely integrable situations, whereas the last one gives us an example of a nonintegrable problem. But each problem has some merits of its own, and it is still a characteristic property of a nonlinear system that a generalization on the procedure is not possible.

Melnikov's equation[96]

Melnikov's equation is written as

$$\frac{\partial u}{\partial t} - \frac{\partial u}{\partial y} + 2k \frac{\partial}{\partial x}|\phi|^2 = 0$$

$$i\frac{\partial \phi}{\partial t} = u\phi + \phi_{xx}$$

(3.6.1)

We start with the usual ansatz:

$$u \sim a_0 \tilde{\psi}^{\alpha}(x,y,t) ; \quad \phi \sim b_0 \tilde{\psi}^B(x,y,t) ,$$

$$\phi^* \sim b_0^* \tilde{\psi}^B(x,y,t) ,$$

where $(x,y,t) = 0$ is the singular manifold. Matching the most singular term leads to

(a) $\quad \alpha = -2 , \quad \beta = -1$

or

(b) $\quad \alpha = -1 , \quad \beta = -1/2 .$

But in case (b), terms left out become more singular, and we get inconsistency. Here in $(2+1)$ dimension, we follow the observation of Kruskal to set

$$\tilde{\psi} = y - \psi(x,t) ,$$

where $\psi(x,t)$ is arbitrary. So we can set

$$u = \sum_{j=0}^{\infty} a_j(t,x)(y - \psi(x,t))^{j-2}$$

$$\phi = \sum_{j=0}^{\infty} b_j(t,x)(y - \psi(x,t))^{j-1}$$

$$\phi = \sum_{j=0}^{\infty} b_j(t,x)(y - \psi(x,t))^{j-1} \, .$$

On the other hand, the leading order coefficients yields

$$2a_0(1+\psi_t) + 2Kb_0 b_0^* \psi_x = 0$$
$$a_0 b_0 + 2b_0 \psi_x^2 = 0$$
$$a_0 b_0^* + 2b_0^* \psi_x^2 = 0 \, ,$$

which yields,

$$a_0 = -2\psi_x^2$$
$$b_0 b_0^* = 1/k(1+\psi_t)\psi_x \, .$$

To search for resonances, we consider next-to-leading-order terms,

$$u = a_0(y - \psi)^{-2} + a_r(y - \psi)^{r-2}$$
$$\phi = b_0(y - \psi)^{-1} + b_r(y - \psi)^{r-2} \, .$$

Similarly for ϕ, substituting in equation (3.6.1), one gets

$$-a_r\{(r-2)\psi_t + (r-2)\} - 2Kb_0 b_0^*(r-2)\psi_x = 2Kb_0^* b_r \psi_x(r-2)$$

$$+ \text{ less singular terms}$$

$$a_r b_0 + a_0 b_r + b_r(r-1)(r-2)\psi_x^2 = \text{ less singular terms}$$

$$a_r b_0 + a_0 b_r + b_r^*(r-1)(r-2)\psi_x^2 = \text{ less singular terms,}$$

whence the determinant of the system matrix turns out to be

$$\begin{vmatrix} -(r-2)(1+\psi_t) & -2Kb_0 \psi_x(r-2) & -2Kb_0^* \psi_x(r-2) \\ b_0 & a_0 + (r-1)(r-2)\psi_x^2 & 0 \\ b_0^* & 0 & a_0 + (r-1)(r-2)\psi_x^2 \end{vmatrix} = \det|T| \, .$$

The roots of $\det|T| = 0$ yields

$$r = -1, 0, 2, 3, 4 .$$

As in the case of equations, in $(1+1)$ dimension resonance at $r = -1$ implies arbitrariness of $\tilde{\psi}$, and hence that of ψ. The resonance at $r = 0$ corresponds to the fact that out of the three coefficients (a_0, b_0, b_0), only two are determined, so at least one is arbitrary. We next consider $r = 1$ and obtain

$$a_1 = -2\psi_{xx}$$

$$b_1 = 1/D\big[a_0 b_0 (1+\psi_t)(i\psi_t - \psi_{xx}) - 2\alpha_0(1+\psi_t)\cdot b_{0x}\psi_x$$

$$+\left\{a_{0t} + 2K\frac{\partial}{\partial x}\left(\frac{\psi_x\psi_t + \psi_x}{K}\right)\right\}\bigg/a_0 b_0$$

$$-i4Kb_0^2\psi_x b_0^*\psi_t - 4kb_0\psi_x^2\big(b_{0x} - b_0^* b_{0x}\big)\Big]$$

$b_1^* = $ complex conjugate of b_1.

Next, $r = 2$ gives

$$b_2 = 1/a_0\left(ib_{0t} - b_{0xx} - a_1 b_1 - a_2 b_0\right)$$

$$b_2 = 1/a_0\left(-ib_{0t} - b_{0xx}^* - a_1 b_1^* - a_2 b_0^*\right)$$

along with a condition

$$\left(a_1\right)_t = 2K\left(b_1 b_0^*\right)_x + 2k\left(b_0 b_1^*\right)_x ,$$

so we have only two equations for three coefficients (a_2, b_2, b_2), and at least one is arbitrary. By a laborious calculation, one can check that the above condition is satisfied identically by the expressions for a_1, b_1, b_1^*. For $r = 3$, we get

$$-a_3(1+\psi_t) - 2Kb_3 b_0^*\psi_x - 2Kb_0 b_3^*\psi_x = -(a_2)_t - 2K(b_2)_x b_0^*$$

$$-2K(b_0)_x b_2^* - 2Kb_2\left(b_0^*\right)_x - 2Kb_1\left(b_1^*\right)_x - 2Kb_0\left(b_2^*\right)_x - 2K(b_1)_x b_{1x}^*$$

$$+2Kb_2 b_1^*\psi_x - 2Kb_1 b_2^*\psi_x \tag{3.6.2}$$

$$a_3 b_0 = i(b_1)_t - ib_2 \psi_t - (b_1)_{xx} + 2(b_2)_{xx} + b_2 \psi_{xx} - a_2 b_1 - b_2 a_1 \tag{3.6.3}$$

$$a_3 b_0^* = -i(b_1^*)_t + ib_2^* \psi_t - (b_1^*)_{xx} + 2(b_2^*)_x \psi_x + b_2^* \psi_{xx} - a_2 b_1^* - b_2^* a_1 . \tag{3.6.4}$$

It can be easily checked that the right side of equations (3.6.3) and (3.6.4) are identical. One equation is left, so one of the coefficients (a_3, b_3, b_3) is arbitrary. The same situation repeats for the resonance at $r = 4$. In total, we have five resonances and at least five arbitrary coefficients, which is in conformity with the Cauchy-Kowalevski condition.

In the next stage, we invoke the similarity reduction of our system to reduce the number of independent variables and then check for the complete integrability. Using the usual Lie point symmetry, one can ascertain that equation (3.6.1) has solutions of the form,

$$u = y^{-1} f\left(x^2/y, t/y\right); \quad z = x^2/y, \quad s = t/y$$

$$\phi = y^{-3/4} g\left(x^2/y, t/y\right) .$$

Then equation (3.6.1) reduces to

$$\frac{\partial f}{\partial s} + f + z \frac{\partial f}{\partial s} + 4K\sqrt{z} \cdot \left(g^* \frac{\partial g}{\partial z} + g \frac{\partial g^*}{\partial z}\right) + S \frac{\partial f}{\partial s} = 0$$

$$i \frac{\partial g}{\partial s} = fg + 2 \frac{\partial g}{\partial z} + 4z \frac{\partial^2 g}{\partial z^2} \tag{3.6.5}$$

$$-i \frac{\partial g^*}{\partial s} = fg^* + 2 \frac{\partial g^*}{\partial z} + 4z \frac{\partial^2 g^*}{\partial z^2} .$$

In the second stage of reduction, we at once observe that equation (3.6.5) admits solutions of the form

$$f = z^{-1} F(s/z), \quad t = s/z$$

$$g = z^{-3/4} G(s/z), \quad g = z^{-3/4} G(s/z) ,$$

whence we get

$$\frac{\partial F}{\partial t} - \sigma KGG^* - 4K\left(G\frac{\partial G^*}{\partial t} + G^*\frac{\partial G}{\partial t} \right) = 0$$

$$4i\frac{\partial G}{\partial t} = 4FG + 15G + 48\cdot t\frac{\partial G}{\partial t} + 16t^2\frac{\partial^2 G}{\partial t^2}$$

$$-4i\frac{\partial G^*}{\partial t} = 4FG^* + 15G^* + 48t\frac{\partial G^*}{\partial t} + 16t^2\frac{\partial^2 G}{\partial t^2} \; ,$$

which is a set of coupled ordinary differential equations. If we now follow the standard procedure, we get resonances at

$$r = -1, 0, 2, 3, 4 \; ,$$

giving rise to the same set of resonances. In our previous discussions, we observed that there is an intimate relation between Hirota's approach and the Painleve analysis. Here also we can obtain exact soliton-like solutions by adopting such an approach. Let us set

$$u = 2\frac{\partial^2}{\partial x^2}\log \tilde{F}, \quad \phi = \tilde{G}/\tilde{F}, \quad \phi^* = \tilde{G}^*/\tilde{F} \; ,$$

so that equations (3.6.1) are transformed to the form

$$\left(D_t - D_y\right)\tilde{F}_x\tilde{F} = -K\tilde{G}^*\tilde{G}$$

$$\left(iD_t - D_x^2\right)\tilde{G}\cdot\tilde{F} = 0$$

$$\left(D_t + D_x^2\right)\tilde{G}^*\tilde{F} = 0 \; .$$

We then set

$$\tilde{F} = 1 + \varepsilon f_1 + \varepsilon^2 f_2 + \dots$$

$$\tilde{G} = \varepsilon g_1 + \varepsilon^2 g_2 + \dots$$

$$\tilde{G} = \varepsilon g_1^* + \varepsilon^2 g_2^* + \dots$$

and compare coefficients of same power of ε. These simple equations immediately yield

$$f_1 = e^{(px+Qy-\Omega t)}, \qquad \begin{aligned} g_1 &= be^{(Lx+My-Nt)} \\ g_1 &= b^* e^{(L^*x+M^*y-N^*t)} \end{aligned},$$

along with $\Omega = Q$, $L^2 = -iN$, all the constants are complex. Also, at the order of ε^2,

$$f_2 = C_2 e^{(L+L^*)x+(M+M^*)y-(N+N^*)t}$$

$$g_2 = d_2 e^{(P+L)x+(Q+M)y-(\Omega+N)t}$$

$$g_2 = d_2 e^{(P^*+L^*)x+(Q^*+M^*)y=(\Omega^*+N^*)t},$$

where

$$c_2 = \frac{Kbb^*}{(L+L^*)\{(N+N^*)-(M+M^*)\}}$$

$$d_2 = \frac{ib(\Omega-N)-b(L-P)^2}{i(\Omega+N)+(P+L)^2}$$

$$d_2 = \frac{ib^*(\Omega-N^*)-b(L^*-P)^2}{i(\Omega^*+N^*)+(P^*+L^*)^2}.$$

Finally, the one- and two-soliton solutions can be written as

$$\phi_1 = g/GF = g_1/(1+f_1); \qquad \phi_2 = (g_1+g_2)/(1+f_1+f_2),$$

whence

$$u = 2\frac{\partial^2}{\partial x^2}(\log F) = \frac{P^2}{2}\mathrm{Sech}^2\left(\frac{Px+Qy+Rt}{2}\right).$$

The present example simply illustrates the effectiveness of the Painleve test for equations in $(2+1)$ dimension.

It also shows that the similarity-reduced equations have the same property. On the other hand, in some special situations, it may be possible to deduce some important relations from such analysis. This is the case with the Burger equation in $(3+1)$ dimension, where we will show that an extended form of Cole-Hopf transformation can be deduced, which linearizes the equation.

Burger equation in (3+1) dimension

An interesting example of a nonlinear PDE was considered by Webb and Zank.[55] These authors considered the Burger equation in (3+1) dimension. Such an equation is written as

$$\left(u_t + u u_x - u_{xx}\right)_x + u_{yy} + u_{zz} = 0 .$$ (3.6.6)

This equation is also known as Zabolotskaya-Khoklov equation, which occurs in nonlinear acoustics. Lie symmetry of this equation was studied by Roy Chowdhury and Naskar.[87] The solution of such an equation represents shocks, as in the case of the usual Burger equation. It also describes propagation of a shock front in magnetogasdynamics. The transverse Laplacian $u_{yy} + u_{zz}$ represents wave diffraction effects.

The Painleve analysis starts with the usual substitution:

$$u(x,y,tz) = \phi^{-p} \sum_{j=0}^{\infty} u_j(x,y,z,t)\phi^j ,$$

where $p = 1$. By balancing terms of order ϕ^{j-4}, we obtain the recurrence relation

$$u_{j-3tx} + (j-3)\left[\phi_x u_{j-2,t} + u_{j-2,x}, \ \phi_t + u_{j-2}\phi_{xt}\right]$$

$$+ (j-2)(j-3)\phi_x\phi_t u_{j-1} + \sum_{k=0}^{j}\left\{(k-k-1)\phi_x^2 u_k u_{j-k}\right.$$

$$+ 2(j-k-2)\phi_x u_{k,x} u_{j-k-1} + u_{k,x} u_{j-k-2,x}$$

$$+ (k-1)(1-2)\phi_x^2 u_k u_{j-k} + (k-1)u_{j-k-1}\left[\phi_{xx} u_k + 2\phi_x u_{kx}\right]$$

$$+ u_{j-k-2}u_{k,xx}\right\} - \left\{(j-1)(j-2)(j-3)\phi_x^3 u_j + (j-2)(j-3)\left[3\phi_x\phi_{xx}u_{j-1}\right.\right.$$

$$+ 3\phi_x^2 u_{j-1,x}\right] + (j-3)\left[\phi_{3x}u_{j-2} + 3\phi_{xx}u_{j-2,x} + 3\phi_x u_{j-2,xx}\right]$$

$$+ u_{j-2,xxx}\right\} + (j-2)(j-3)\left(\phi_y^2 + \phi_z^2\right)u_{j-1} + (j-3)\left[2\phi_y u_{j-2,y} + -\phi_z u_{j-2,z}\right.$$

$$\left.+ u_{j-2}\left(\phi_{yy} + \phi_{zz}\right)\right] + u_{j-3,yy} + u_{j-3,zz} = 0 .$$

For $j = 0$, one gets

$u_0 = -2\phi_x$. Resonances are seen to occur at $j = -1, 2, 3$.

$$j = 1, \quad u_1 = \frac{\phi_{xx} - \phi_t}{\phi_x} - \frac{\phi_y^2 + \phi_z^2}{\phi_x^2} \tag{3.6.7}$$

$$j = 2, \quad \phi_{xx}\left(\phi_y^2 + \phi_z^2\right) - 2\phi_x\left(\phi_y\phi_{xy} + \phi_z\phi_{xz}\right) + \phi_x^2\left(\phi_{yy} + \phi_{zz}\right) = 0$$

$$j = 3, \quad u_{0,tx} + \left(u_0 u_1\right)_{xx} - u_{0,xxx} + u_{0,yy} + u_{0,zz} = 0 \; .$$

Resonance condition at the order $j = 3$ is automatically satisfied when ϕ satisfies second equation of (3.6.7). Because this is not identically satisfied we say that the Burger equation in $(3+1)$ dimension is conditionally integrable and does not possess the full Painleve property. But a consistent truncation can be obtained. The equation at $j = 4$ gives

$$\left(u_{1,t} + u_1 u_{1,x} - u_{1,xx}\right)_x + u_{1,yy} + u_{1,zz} = 0 \; , \tag{3.6.8}$$

and

$$u = -2\phi_x/\phi + u_1 \; ,$$

where u_1 is known from (3.6.7) in terms of ϕ. So essentially we have two conditions on ϕ. From the general theory of PDE we know that this quasilinear equation for ϕ can be linearized if we consider $X(y,z,\phi,t)$ as the new dependent variable. It turns out that we get

$$X_{yy} + X_{zz} = 0 \; ,$$

with general solution

$$X = \alpha(\xi,\phi,t) + \beta(\eta,\phi,t)$$

$$\xi = y + iz , \quad \eta = y - iz \; .$$

α, β are arbitrary analytic functions, so equation (3.6.7) gives

$$u_1 = -X_{\phi\phi}/X_\phi^2 + X_t - 4X_\xi X_\eta \; , \tag{3.6.8a}$$

and the equation (3.6.8) for u_1 becomes

$$X_\phi^8 \left\{ (u_{1t} + u_1 u_{1x} - u_{1xx})_x + u_{1yy} + u_{1zz} \right\}$$

$$= X_\phi^8 \left(-16 X_{\xi\xi} X_{\eta\eta} \right) + X_\phi^7 \left\{ X_{tt\phi} - 4 \left(X_\xi X_\eta \right)_{\phi t} \right.$$

$$\left. -4 \left(X_\eta X_{\xi t\phi} + X_\xi X_{\eta t\phi} \right) + 16 \left[X_\eta \left(X_{\xi\xi} X_\eta \right)_\phi + X_\xi \left(X_{\eta\eta} X \right)_\phi \right] \right\}$$

$$+ X_\phi^5 \left\{ -2 X_{\phi\phi\phi t} + 4 \left(X_\eta X_\xi \right)_{\phi\phi\phi} + 8 \left(X_{\xi\phi} X_{\eta\phi\phi} + X_{\eta\phi} X_{\xi\phi\phi} \right) \right.$$

$$\left. +4 \left(X_\eta X_{\xi\phi\phi\phi} + X_\xi X_{\eta\phi\phi\phi} \right) \right\} + X_\phi^4 \left\{ 6 X_{\phi\phi} X_{\phi\phi t} + 2 X_{\phi\phi\phi} X_{\phi t} \right.$$

$$\left. -8 X_{\phi\phi} \left[3 \left(X_\xi X_\eta \right)_{\phi\phi} - X_{\xi\phi} X_{\eta\phi} \right] - 8 X_{3\phi} \left(X_\eta X_\xi \right)_\phi \right\}$$

$$+ X_\phi^3 \left\{ -6 X_{\phi\phi}^2 X_{\phi t} + X_{\phi\phi\phi\phi\phi} + 24 X_{\phi\phi}^2 \left(X_\eta X_\xi \right)_\phi \right\}$$

$$+ X_\phi^2 \left\{ -10 X_{\phi\phi} X_{\phi\phi\phi\phi} - 6 X_{\phi\phi\phi}^2 \right\} + 48 X_\phi X_{\phi\phi}^2 X_{\phi\phi\phi} - 36 X_{\phi\phi}^4 = 0 \ . \tag{3.6.9}$$

It is now convenient to search for the solution of X in the form

$$X = \omega(y, z, t) = h(\phi, t) \ ,$$

where ω satisfies the Laplace equation

$$\omega_{yy} + \omega_{zz} = 4\omega_{\xi\eta} = 0 \ .$$

Equation (3.6.9) then leads to the following form of,

$$\omega = a\xi^2 + b\xi + C + a^* \eta^2 + b^* \eta + C^*$$

$$= 2 \left[a_R \left(y^2 - z^2 \right) - 2a_1 y_z + b_R y - b_1 z + C_R \right] , \tag{3.6.9a}$$

where we have separated out the real and imaginary past of each constant $a = a_R + i a_i$ and so on. Equation (3.6.9) then reduces to

$$-64|a|^2 h_\phi^4 + h_\phi^3 h_{tt\phi} - 2h_\phi h_{\phi\phi t} + 6h_{\phi\phi}h_{\phi\phi t} + 2h_{\phi\phi\phi}h_{\phi t}$$

$$-6h_{\phi\phi}^2 \frac{h_{\phi t}}{h_\phi} + \frac{\partial^2}{\partial\phi^2}\{h,\phi\} - 5\frac{h_{\phi\phi}}{h_\phi}\frac{\partial}{\partial\phi}\{h,\phi\}$$

$$+5\{h_{\phi\phi}/h_\phi\}^2\{h,\phi\} - 2\{h,\phi\}^2 = 0 \ , \tag{3.6.10}$$

where

$$\{h,\phi\} = \frac{\partial}{\partial\phi}\frac{h_{\phi\phi}}{h_\phi} - \frac{1}{2}\left(\frac{h_{\phi\phi}}{h_\phi}\right)^2 \ .$$

By regarding ϕ as a function of h and t, equation (3.6.10) reduces in a remarkable way to

$$\frac{\partial}{\partial h}\left(\rho_t + \rho\rho_\eta - \rho_{hh} - 64|a|^2 h\right) = 0 \ , \tag{3.6.11}$$

where

$$\rho = 1/\phi_h\left(\phi_{hh} - \phi_t\right) \ .$$

If we choose $|a| = 0$, then this gives back the one-dimensional Burger equation ρ. Using the standard Cole-Hopf transformation

$$\rho = -\frac{2\psi_h}{\psi} \ ,$$

equation (3.6.11) is transformed into

$$\frac{\partial^2}{\partial h^2}\left(\frac{\psi_t - \psi_{hh}}{\psi} + 16|a|^2 h^2\right) = 0 \ . \tag{3.6.12}$$

Integrating with respect to h,

$$\psi_t - \psi_{hh} + \left[16|a|^2 h^2 + p(t)h + q(t)\right] = 0 \ .$$

Equation (3.6.11) and (3.6.12) gives an equation connecting ϕ and ψ

$$\phi_t - \phi_{hh} - 2\frac{\psi_h}{\psi}\phi_h = 0 \ ,$$

so the Painleve expansion and the structure of u_1 given by (3.6.8a) leads to

$$u = -2\frac{\phi_h}{\phi} + u_1$$

$$u_1 = -\frac{2\psi_h}{\psi} + \omega_t - \left(\omega y^2 + \omega_z^2\right) ,$$

where ω is given as

$$\omega = 2\left[a_R\left(y^2 - z^2\right) - 2a_1 y_z + b_R y - b_1 z + C_R\right] .$$

It is also interesting to note that

$$h = x - \omega(y,z,t) .$$

At this point it may be noted that there are other ways to get the solution of equation (3.6.9). In fact, we can also use the ansatz:

$$X = v(\xi) f(\phi,t) + h(\phi,t) ,$$

where $v(\xi)$ is an arbitrary function of (ξ). Note that we have $X_{\xi\eta} = 0$ as required above.

Now

$$X_\phi = v(\xi) f_\phi + h_\phi = f_\phi[v(\xi) + g] ; \quad g = h_\phi/f_\phi .$$

If we substitute this in equation (3.6.9) and equate various powers of X_ϕ we get

$$-36\left(g_\phi f_\phi\right)^4 = 0 ,$$

from which we infer that $f_\phi \neq 0$, then $g_\phi = 0$. Other conditions are

$$\frac{\partial^2}{\partial\phi^2}\{f,\phi\} - 5\frac{f_{\phi\phi}}{f_\phi}\frac{d}{d\phi}\{f,\phi\} + 5\left\{\frac{f_{\phi\phi}}{f_\phi}\right\}^2 \{f,\phi\}$$

$$-2\{f,\phi\}^2 - f_\phi^4 \frac{\partial^2}{\partial f^2}\{\phi,f\} = 0 \qquad\qquad (3.6.13a)$$

$$\frac{\partial}{\partial t}\{f,\phi\} = 0 \tag{3.6.13b}$$

$$\frac{\partial}{\partial t}\left(g_t f_\phi^2\right) = 0 \tag{3.6.13c}$$

$$f_{tt\phi} = 0 \ .$$

It is now simple to solve these subsidiary equations:

(a) Equations (3.6.13a) to (3.6.13c) have the solutions

$$g(t) = C_2 + \frac{C_1}{t+c}$$

$$f(t) = A(\phi)(t+c) + m(t)$$

$$h(t) = g(t)f + p(t)$$

along with

$$\frac{d^2}{dA^2}\{\phi, A\} = 0 \ ,$$

which can be integrated at once to yield

$$\{\phi, A\} = 2(K_1 A + K_2) \ .$$

Next one may observe that ϕ has a solution of the form $\phi = Y_1/Y_2$, where Y_1 and Y_2 are independent solutions of

$$Y''(A) + (K_1 A + K_2)Y = 0 \ .$$

The general solution of this equation is

$$Y = (\Delta A)^{1/2}\left[a J_{1/2}\left(2/3 K_1^{1/2}(\Delta A)^{1/2}\right) + b Y_{1/3}\left(2/3 K_1^{1/2}(\Delta A)^{3/2}\right)\right]$$

a, b are arbitrary constants

$$\Delta A = A + K_2 / K_1 .$$

In the above expressions, $J_{1/3}$, $Y_{1/3}$ are Bessel functions of the first and second kind.

(b) Equations (3.6.13a) to (3.6.13c) also have solutions

$$g(t) = d_1 t + d_2$$
$$f(t) = A(\phi) + m(t)$$
$$h(t) = g(t) f + p(t)$$

and can follow the same analysis as described above. In the above discussions, we have shown that the Burger equation in $(3+1)$ dimension is conditionally integrable, and yet one can analyze the various solutions through the Painleve test. The important point to notice is that the Painleve test can be used to generate useful information even though the system is not completely integrable. We have already discussed such situations in Section 3.6, where we studied the Painleve analysis for the nonintegrable systems, but in $(1+1)$ dimension. In the course of our analysis, we have deduced the generalized Cole-Hopf transformation:

$$u = -\frac{2\psi_n}{\psi} + \omega_t - \left(\omega y^2 + \omega_z^2 \right)$$
$$h = x - \omega(y, zt)$$

where ω is given in equation (3.6.9a). This transformation gives us a mapping of the Burger equation in $(3+1)$ dimension to

$$\psi_t - \psi_{hh} + \left[16 |a|^2 h^2 + p(t) h + q(t) \right] \psi = 0 ,$$

the generalized heat conduction equation.

Nonlinear Schrodinger in (2+1) dimension

One of the most important nonlinear equations that has been discussed and used in various fields of physics is the Nonlinear Schrodinger in $(2+1)$ dimension. Unlike the Burger equation in $(3+1)$ dimension, it is completely integrable. Here we discuss how the Painleve test can be implemented in such a situation.

The equations under consideration can be written as[47,48]

$$p_t = Ap_{xx} - Bp_{yy} + (r-s)p$$
$$q_t = Bq_{yy} - Aq_{xx} + (s-r)q$$
$$r_x = 2B(pq)y \qquad\qquad\qquad (3.6.14)$$
$$s_y = 2A(pq)_x \ .$$

For the leading order analysis, we set $p \sim p_0\phi^\alpha$, $q \sim q_0\phi^\beta$, $r \sim r_0\phi^\gamma$, and $s \sim s_0\phi^\delta$, where the singularity manifold ϕ is chosen to be $\phi = x - f(y,t)$ and (p_0, q_0, r_0, s_0) are all functions of (y,t). Similar assumptions were also used by Jimbo et al., Goldstein, and Infeld.[49] Substituting these in equation (3.6.14) and comparing the most singular terms, we get

$$\alpha = -1, \quad \beta = -1, \quad \gamma = \delta = -2 \ .$$

But to get an idea about the coefficients of the leading terms, we write these out in complete detail. These are

(a) $p_{0t}\phi^\alpha - \alpha p_0\phi^{\alpha-1}f_t$
$$= Ap_0\alpha(\alpha-1)\phi^{\alpha-2} - Bp_{0yy}\phi^\alpha + 2Bp_{0y}\alpha\phi^{\alpha-1}f_y$$
$$- Bp_0\alpha(\alpha-1)\phi^{\alpha-2}f_y^2 + Bp_0\alpha\phi^{\alpha-1}f_{yy} + r_0p_0\phi^{\alpha+r} - s_0p_0\phi^{\alpha+\delta}$$

(b) $q_{0t}\phi^\beta - \beta q_0\phi^{\beta-1}f_t = Bq_0\beta(\beta-1)\phi^{\beta-2}f_y^2 - Aq_0\beta(\beta-1)\phi^{\beta-2}$
$$+ Bq_{0yy}\phi^\beta - 2Bq_{0y}\phi^{\beta-1}f_y - Bq_0\beta/\beta^{-1}f_{yy} + q_0s_0\phi^{\beta+\delta} - q_0r_0\phi^{\beta+\gamma}$$

(c) $r_0r\phi^{r-1} = 2B(p_0q_0)_y\phi^{\alpha+\beta} - 2B(p_0q_0)(\alpha+\beta)\phi^{\alpha+\beta-1}f_y$

(d) $s_{0y}\phi^\delta - s_0\delta\phi^{\delta-1}fy = 2A(p_0q_0)(\alpha+\beta)\phi^{\alpha+\beta-1} \ .$

Most singular terms lead to

$$p_0q_0 = -fy; \quad r_0 = 2Bf_y^2; \quad s_0 = 2A \ .$$

Note that we may have other possibilities for the exponents. For example, $\delta = r = -1$, but $\alpha + \beta = -1$. But in these cases, the leading terms are almost decoupled

$$Ap_{xx} - Bp_{yy} = 0, \qquad Bq_{yy} - Aq_{xx} = 0$$

$$r_x = 2B(pq)_y, \qquad s_y = 2A(pq)_x.$$

So we do not consider this. Other possibilities are $\delta = -2$, $r = -2$, $\alpha + \beta = -2$, but $\alpha \neq -1$ and $\beta \neq -1$. We have fractional values of (α, β) so that $(\alpha + \beta) = -2$. Due to the fractional nature of the exponents, we also leave this situation.

At the next stage we determine the resonance positions by setting

$$p = \sum_{j=0}^{\infty} p_j \phi^{j-1}; \qquad q = \sum_{j=0}^{\infty} q_j \phi^{j-1}$$

$$r = \sum_{j=0}^{\infty} r_j \phi^{j-2}; \qquad s = \sum_{j=0}^{\infty} s_i \phi^{j-2}.$$

Equating coefficients of ϕ^{m-3}, we get

$$
\begin{pmatrix}
R & 0 & p_0 & -p_0 \\
0 & R_1 & -q_0 & q_0 \\
2Bf_y q_0(m-2) & 2Bf_y p_0(m-2) & m-2 & 0 \\
2Aq_0(m-2) & 2Ap_0(m-2) & 0 & (m-2)f_y
\end{pmatrix}
\begin{pmatrix}
p_m \\ q_m \\ r_m \\ s_m
\end{pmatrix}
=
\begin{pmatrix}
X \\ Y \\ Z \\ T
\end{pmatrix}
\qquad (3.6.15)
$$

where

$$X = p_{m-2,t} - p_{m-1}(m-2)f_t + Bp_{m-2,yy} - 2Bp_{m-1,y}(m-2)f_y$$

$$-Bp_{m-1}(m-2)f_{yy} + \sum_{n=1}^{m-1} r_{m-n}p_n - \sum_{n=1}^{m-1} s_{m-n}p_n$$

$$Y = q_{m-2,t} - q_{m-1}(m-2)f_t - Bq_{m-2,yy} + 2Bq_{m-1,y}(m-2)f_y$$

$$+Bq_{m-1}(m-2)f_{yy} - \sum_{n=1}^{m-1} s_{m-n}q_n + \sum_{n=1}^{m-1} r_{m-n}q_n$$

$$Z = 2B \sum_{n=0}^{m-1} (p_{m-n-1}q_n)_y - 2Bf_y \sum_{n=1}^{m-1} p_{m-n}q_n(m-2)$$

$$T = S_{m-1,y} - 2A \sum_{n=1}^{m-1} p_{m-n} q_n (m-2)$$

$$R = A(m-1)(m-2) + r_0 - s_0 - B(m-1)(m-2) f_y^2$$

$$R_1 = -A(m-1)(m-2) - r_0 + s_0 + B(m-1)(m-2) f_y^2 .$$

Resonance positions are those values at $m = r$ for which the determinant of the matrix on the left side of equation (3.6.15) is 0. We find

$$\det[\ \ ] = m(m+1)(m-2)^2 (m-3)(m-4) ,$$

so we get resonances at

$$r = 0, -1, 2, 2, 3, 4 .$$

Expansion coefficients
$r = -1$ corresponds to the arbitrariness of

$$\phi = x - f(y,t)$$

$r = 0$ represents the arbitrariness of either p_0 or q_0.
$r = 1$. We get from equation (3.6.15)

$$p_1 = 1/(r_0 - s_0)\left(-Bp_0 f_{yy} + q_0 f_t + 2Bp_{0y} f_y\right)$$
$$q_1 = 1/(r_0 - s_0)\left(-Bq_0 f_{yy} - q_0 f_t + 2Bq_{0y} f_y\right)$$
$$r_1 = 2Bf_{yy}$$
$$s_1 = 0$$

$r = 2$ (double resonance). In this case,

$$p_2 = 1/(r_0 - s_0)\left[-p_0(r_2 - s_2) + p_{0t} + Bp_{0yy} - r_1 s_1\right]$$
$$q_2 = 1/(r_0 - s_0)\left[-q_0(r_2 - s_2) - q_{0t} + Bq_{0yy} - r_1 s_1\right]$$

r_2, s_2 arbitrary,

subject to the condition

$$(p_0 q_{1t} + p_1 q_0)_y = 0, \quad s_{1y} = 0 .$$

One can at once check that these conditions are immediately satisfied by the above expressions.

$r = 3$. Here we obtain

$$r_3 = 2B(p_2 q_0 + p_1 q_1 + p_0 q_2)_y - 2Bf_y(p_3 q_0 + p_2 q_1 + p_1 q_2 + p_0 q_3)$$

$$s_3 = 1/f_y \left[s_{2y} - 2A(p_3 q_0 + p_2 q_1 + p_1 q_2 + p_0 q_3) \right] (q_0 p_3 + p_0 q_3)$$
$$= f_y/(r_0 - s_0) \left\{ -1/f_y (r_0 - s_0)(p_2 q_1 + q_2 p_1) \right.$$
$$+ 2B(p_2 q_0 + p_1 q_1 + p_0 q_0)_y - s_{2y}/f_y - 1/p_0 \left[p_{1t} - p_2 f_t + B p_{1yy} \right.$$
$$\left. -2Bp_{2y}f_y - Bp_2 f_{yy} + s_2 q_1 - r_2 q_1 - r_1 p_2 \right] \right\} .$$

Therefore, one of the coefficients p_3 and q_3 can be assumed to be arbitrary, subject to the following compatibility condition

$$2B_f \left(p_0 q_{2y} - q_0 p_2 \right) + 3B f_{yy} (p_0 q_2 - q_0 p_2) - (p_0 q_2 + q_0 p_2) f_y$$
$$+ (p_0 q_1 - q_0 p_1)(r_2 - s_2) + (s_0 p_{1t} + p_0 q_{1t}) + B(q_0 p_{1yy} - p_0 q_{1yy})$$
$$+ s_1 (q_0 p_2 - p_0 q_2) = 0 .$$

By inserting the values of (p_0, q_0), (p_2, q_2), etc. in the above equation, one observes that it is identically satisfied.

Next, consider the situation at $r = 4$. At the last resonance position, the recurrence relation leads to

$$r_4 = 1/2 \left[-4Bf_y(q_0 p_4 + p_0 q_4) + 2B(p_3 q_0 + p_2 q_1 + p_1 q_2 + p_0 q_3)_y \right.$$
$$\left. -4Bf_y(p_3 q_1 + p_2 q_2 + p_1 q_3) \right]$$

$$s_4 = 1/2 f_y \left[-4A(q_0 p_4 + p_0 q_4) + s_{3y} - 4A(q_3 p_1 + p_2 q_2 + p_3 q_1) \right] (q_0 p_4 - p_0 q_4)$$
$$= 1/(s_0 - r_0) \left\{ f_y \left[B(p_3 q_0 + p_2 q_1 + p_1 q_2 + q_0 p_3)_y - 2Bf_y (p_3 q_1 + p_2 q_2 + p_1 q_3) \right] \right.$$
$$\left. - \left[\tfrac{1}{2} s_{3y} - 2A(p_3 q_1 + p_2 q_2 + p_1 q_3) \right] \right\} ,$$

subject to a compatibility condition, which can be written as follows:

$$
\begin{aligned}
&\left(p_0 q_{2t} - q_0 p_{2t}\right) - 2\left(p_0 q_3 - q_0 p_3\right)f_t - B\left(p_0 q_{2yy} + q_0 p_{2yy}\right) \\
&+4B\left(p_0 q_{3y} + q_0 p_{3y}\right)f_y + 2B\left(q_0 p_3 + p_0 q_3\right)f_{yy} + r_3\left(p_0 q_1 + p_1 q_0\right) \\
&+r_2\left(p_0 q_2 + q_0 p_2\right) + r_1\left(p_0 q_3 + p_3 q_0\right) - s_3\left(p_0 q_1 + q_0 p_1\right) \\
&-s_2\left(p_0 q_2 + p_2 q_0\right) - s_1\left(p_0 q_3 + q_0 p_3\right) - 2\Big\{ f_y \Big[B\left(p_3 q_0 + p_2 q_1 + p_1 q_2 + p_0 q_3\right)_y \\
&-2B f_y\left(p_3 q_1 + p_2 q_2 + p_1 q_3\right) \Big] - \Big[1/2 s_{3y} - 2A\left(p_3 q_1 + p_2 q_2 + p_1 q_3\right) \Big] \Big\} = 0 \ .
\end{aligned}
$$

It is astonishing that such a complicated-looking condition is also identically satisfied by the coefficients p_3, q_3, p_1, q_1, etc. So we have a total of six resonances and six arbitrary constants, which conforms with the Cauchy-Kovalevski condition. One can now truncate the series expansion and check whether the compatibility relations are in order. A Backlund transformation of the form

$$
\begin{aligned}
p &= p_0/\phi + p_1 \\
q &= q_0/\phi + q_1 \\
r &= r_0/\phi^2 + r_1/\phi + r_2 \\
s &= s_0/\phi^2 + r_1/\phi + s_2
\end{aligned}
\qquad (3.6.16)
$$

can be constructed, where the set $\left(p_1, q_1, r_2, s_2\right)$ satisfy the same set of equations. Such transformations can also be effectively used to generate various soliton solutions. We do not go into these details and leave this as an exercise for the reader.

3.7 Negative resonances and Painleve expansion

From our previous analysis, we have seen that the Painleve test has been proved to be quite a successful tool for isolating integrable nonlinear ODEs and PDEs. But in various applications, people have faced situations where negative resonances appear. Examples of such situations can be found in discussions on special equations (Chapter 3, see 3.5.7) and nonlinear equations in more than one dimension (Chapter 3, see 3.6). Perhaps the worst situation is given by Chazy's equation, where all the resonance positions are negative because the expansion used in the Painleve test involves only summation over positive indices. It appears that negative resonance reduces the number of required arbitrariness demanded by the test. In this connection, an interesting set of works were done by Fordy and his collaborators, who showed that negative resonances can contain important information regarding the integrability of the equation. It would be quite erroneous to neglect them.

Another important aspect of references (61) is that they showed that all the branches of the Painleve analysis may be equally important. The basic technique of Conte, Fordy and Pickering is to simultaneously test the nonlinear equation and its linearization, treated as a coupled system. For the linearized equation, the Painleve analysis reduces to a Fuchsian analysis about a regular singularity. The roots of the indicial equation of the linearized problem are just the resonances of the nonlinear system.

For convenience, we consider the following situation:

$$u_t = K[u] \, , \tag{3.7.1}$$

where $K[u]$ is a polynomial function of u, $u_x, \ldots, u_{Nx}$. The linearized version of (3.7.1) is

$$\omega_t = K'[u]\omega$$
$$K'[u] = d / d\varepsilon K[u + \varepsilon W]\big|_{\varepsilon=0} \tag{3.7.2}$$

We follow the conformal invariant form of Painleve expansion introduced by Conte and discussed in Chapter 3. We set

$$u^{(p)} = \psi^{-\alpha} \sum_{i=0}^{\infty} u_i \psi^i \, , \tag{3.7.3}$$

where we have written ψ instead of ϕ. The leading order analysis of the equation gives a number of possible choices corresponding to the various starting terms u_0. If we denote the dominant part of $K[u]$ as $K_d[u]$, then u_0 is the solution of

$$K_d\left[u_0\psi^{-\alpha}\right]\bigg|_{x'-1=u_0'=0} = P_0(u_0)\chi^\beta = 0 \, , \tag{3.7.4}$$

where β is the weight of the dominant expression $\hat{K}$ and P_α a polynomial. The expansion coefficients u_i are determined recursively by

$$\left[K_d' \left[u_0 \psi^{-\alpha} \right] \psi^{i-\alpha} \Big|_{x'=1, u_0'=0} \right] u_i = f(u_0, \ldots, u_{i-1} S, C) \ ,$$ (3.7.5)

where the right side is an expression depending on $u_0, \ldots, u_{i-1}$, SC. The coefficients of u_i on the left vanish for certain values of i, called resonance these "i" values must be "positive" integers for an equation to be integrable. That is, r can have any value $r_i \varepsilon_R^R$;

$$R = \{ r_1, \ldots, r_n \} r_1 < r_2 < \ldots < r_n \ .$$

In the case of a conventional Painleve test, one always assumes that if there are several branches there will be one branch known as the principal branch that will have several resonances equal to the degree of the equation and all the r, s will be positive integers, except one that will be equal to -1.

Fordy and Pickering relaxed these conditions to some extent and used the following modified ones:

1.   Each possible choice of α and all corresponding r_i must be integers, the r_i being distinct.

2.   All branches should be such that at any positive resonance, the compatibility conditions are identically satisfied.

3.   There will be at least one branch with the number of resonances $n = N$, the degree of the operator $K(u)$.

The next step in this modified approach is to consider the linearized equation (3.7.2), with u replaced by the Painleve expansion for each of the branches. For each branch, we can write

$$K_d' \left[u^{(p)} \right] \omega = I' \left[u^{(p)} \right] \omega \ .$$

K_d' and I' are the linearized forms of the dominant, and other parts of $K - u_t$. $K_d' \left[u^{(p)} \right]$ are scaled in such a way that $\psi = 0$ is a regular singularity in the Fuchsion sense. We search for expansion

$$\omega = \psi^\sigma \sum_{i=0}^{\infty} \omega_i \psi^i \ , \qquad \omega_0 \neq 0 \ ,$$

where $\sigma \varepsilon \{\sigma_1,...,\sigma_n\}$ is a root of the indicial equation. Comparing this with equation (3.7.5), we see that

$$\sigma_i = r_i - \alpha \quad \text{for} \quad r_i \varepsilon R(i = 1,...,n) \ .$$

We now require that

1. the indicial equation has n distinct integer solutions

    $$(\sigma_1,...,\sigma_n) \ ;$$

2. the compatibility conditions arising at each σ_i $1 \le i \le n$ be identically satisfied. These are essential for the general solution of the linear equation to be single valued.

Fordy and his collaborators used Chazy's equation to demonstrate the power of this modified form of analysis.

Example:
Chazy's equation is written as

$$K[u] = u_{xxx} - 2uu_{xx} + 3u_x^2 = 0$$

$$K'[u]\omega = \omega_{xxx} - 2u\omega_{xx} + 6u_x\omega_x - 2u_{xx}\omega = 0 \ .$$

(3.7.6)

The two branches are

(i) $\alpha = 1, \ u_0 = -6$

        $K_d[u] = K[u]$

        $\beta = -4 , \ \ R = \{-3,-2,-1\} \ ,$

        all negative resonances.

(ii) $\alpha = 2, \ u_0$ arbitrary

        $K_d[u] = -2uu_{xx} + 3u_x^2$

        $\beta = -6 , \ \ R = \{-1, 0\}.$

Because neither branch contains negative resonances, the usual Painleve test cannot be applied, but the branch (i) has the required number of resonances. The linearized forms are

(i) $\quad \omega_{xxx} - 2u\omega_{xy} + 6u_x\omega_x - 2u_x\omega = 0$

(ii) $\quad -2u\omega_{xx} + 6u_x\omega_x - 2u_{xx}\omega = -\omega_{xxx}$.

In (ii) we have put the less-dominant part on the right side. Using now Fuchsian type expansion for each of these leads to

(i) $\quad (\sigma + 4)(\sigma + 3)(\sigma + 2) = 0$

(ii) $\quad (\sigma + 3)(\sigma + 2) = 0$

as the indicial equations. We may summarize as follows

(i) $\quad u = \psi^{-1} \sum\limits_{i=0}^{\infty} u_i \psi^i , \quad u_0 \neq 0$

$\quad\quad \omega = \psi^{-4} \sum\limits_{i=0}^{\infty} \omega_i \psi^i , \quad \omega_0$ arbitrary

(ii) $\quad u = \psi^{-2} \sum\limits_{i=0}^{\infty} u_i \psi^i$

$\quad\quad \omega = \psi^{-3} \sum\limits_{i=0}^{\infty} \omega_i \psi^i , \quad \omega_0$ arbitrary .

Substituting these forms of ω and u in equations (3.7.1) and (3.7.2), we get

$$K[u] = \psi^{-4} \sum\limits_{i=0}^{\infty} K_i \psi^i = 0$$

$$K'[u] = \psi^{-7} \sum\limits_{i=0}^{\infty} l_i \psi^i = 0 ,$$

where

$\quad\quad K_0 = -u_0(u_0 + 6)$

$\quad\quad K_1\big|_{u_0=-6} - 24u_1$

$$K_2\big|_{u_0=-6} = 60(u_2+S)$$
$$u_1 = 0 \ .$$

Due to the absence of any positive resonances, we immediately get

$$u_0 = -6, \qquad u_1 = 0, \qquad u_2 = -s \ . \tag{3.7.6}$$

One can now try to construct the expansion for u and ω. But at this stage there are some obstructions due to the fact that higher roots of the indicial equation are positive resonances, $s_i = \sigma_i + 4$ corresponding to the negative resonances of u. The following compatibility conditions arise

$$S_1 = 0, \qquad l_0 = -20\omega_0(u_0+6) = 0$$
$$S_2 = 1, \qquad l_1\big|_{u_0=-6} = -40\omega_0\omega_1 = 0$$
$$S_3 = 2, \qquad l_2\big|_{u_0=-6} = -64\omega_0(u_2+S) = 0$$
$$u_1 = 0$$

which are identically satisfied due to equation (3.7.6).

A very similar situation is seen to occur with Bureau's equation. Here

$$K[u] = u_{xxx} - 7uu_{xx} + 11u_x^2 = 0$$
$$K'[u]\omega = \omega_{xxx} - 7u\omega_{xx} + 22u_x\omega_x - 7u_{xx}\omega = 0 \ .$$

There are two branches:

(a) $\alpha = 1, \quad u_0 = -2, \quad K_d[u] = K[u]$
      $\beta = -4, \quad R = \{-6, -1, -1\}$

(b) $\alpha = 7/4, \quad u_0$ arbitrary,
      $K_d[u] = -7uu_{xx} + 11u_x^2$
      $\beta = -11/2, \quad R = \{-1, 0\} \ .$

But here a source of difficulty is the double root -1, which implies that the indicial equation will have a double root at -2. So u will have a logarithm in the expansion, hence we are to modify the expansion of u as

$$u = \psi^{-1} \sum_{i=0}^{\infty} \left(\sum_{j=0}^{i} u_{ij} (\ln \psi)^i \right) \psi^{-1}$$

$u_{00} = -2$, $\quad u_{01}$, $\quad u_{11}$, $\quad u_{00}$ are arbitrary.

In a later communication, Fordy and Pickering generalized the above methodology to include higher-order terms of perturbation instead of only the first-order linearized version. Because this approach may be useful in more complicated situations, we include a short discussion of it below.

Perturbative Painleve approach

In the perturbative Painleve approach, one tries to find a Laurent expansion for any solution that is near to the solution obtained by the standard Painleve approach. This is done by a perturbation in a parameter ε not occurring in the equation. Let us follow the notation of Fordy *et al.* and denote usual Painleve expansion as $\left(u^0, E^0 \right)$ and look for a nearby solution normally represented by an infinite perturbation series in the parameter ε. We set

$$u = \sum_{n=0}^{\infty} \varepsilon^n u^n \quad E \equiv K[u,x] = 0$$

$$E = \sum_{n=0}^{\infty} \varepsilon^n E^n = 0 \ . \tag{3.7.7}$$

The condition that this expansion still must be a solution generates the infinite sequence of successive differential equations,

$$n = 0 : E^{(0)} \equiv K\left[u^0, x \right] = 0$$

$$n \geq 1 : E^n = K_d'\left[u^0, x \right] u^n - R^n\left(u^0, \dots, u^{n-1} \right) = 0 \ . \tag{3.7.8}$$

$R^{(n)}$ represents the contribution of previous terms of the expansion, and R^1 is identically 0. Each higher-order equation is a linear inhomogeneous differential equation whose left side is the Frechet operator $K_d'\left[u^0, x \right]$ acting on u^n. As we have already discussed in Section 3.7, when the Painleve series u^0 is substituted into the linearized equation,

$$K'_d\left[u^0, x\right]u' = 0 ,$$

the resulting equation for u' is of Fuchsian type, its indices being $i+p$, where i runs over the Painleve resonances. At first order, an arbitrary coefficient is introduced at each index, but one should note that not all of these are new. But because the equation for u' is linear, the usual Fuchs-Frobenius theory holds. For each solution σ' of the indicial equation, there is a solution u' with leading behavior ψ^0. If any σ is noninteger, then the solution exhibits branching. After first order, the function u^n satisfies an inhomogeneous, linear differential equation.

The indicial equation is the same for all $n \geq 1$, but for $n \geq 2$ the leading order behavior of u^n is determined by the singularity order of the right side.

$$u^n = \sum_{j=n\rho}^{\infty} u^n, \ \psi^{j+p}$$

$$E^n = \sum_{j=n\rho}^{\infty} E_j^n, \ \chi^{j+q} \tag{3.7.9}$$

$$u = \sum_{n=0}^{\infty} \varepsilon^n \left(\sum_{j=n\rho}^{\infty} u_j^n \psi^{j+p} \right) = \sum_{-\infty}^{\infty} u_h \psi^{j+p}$$

where ρ denotes the smallest index of the nonlinear system (3.7.7), an integer lower than or equal to -1. Coefficients u_j^n satisfy linear algebraic equation just like the unperturbed case

$$E_j^n = \rho(j)u_j^n + Q_j^n = 0 .$$

The form of $P(j)$ and Q_j are to be determined separately in each case. We set

$$u = (u_1, \ldots, u_N), \quad p = (p_1, \ldots, p_N) ,$$

then

$$u_j \sim \psi^{p_j} \quad \text{and} \quad u_j \psi^{j+p} = \left(u_{1j} \psi^{j+p_1}, \ldots, u_N \psi^{j+p_N} \right) .$$

Each set of dominant terms provides different sets of leading behaviors. As solutions of the algebraic equation,

$$K_d\left[u_0\psi^P, x\right]\Big|_{\text{grad } u_j=0,\ \text{grad}\psi=A_0} \equiv F(u_0, x, A_0)\psi^q = 0 ,$$

A_0 is constant, and F is a polynomial in u_0. Each pair (p, u_0) defines an expansion family or branch. But the reader should be very careful not to confuse this with the branching of a solution. For each branch, the other coefficients u_j are determined recursively by the algebraic equation,

$$j \geq 1:\ E_j = P(j)u_j + Q_j = 0$$

$$Q_j = \text{a column vector depending on } \left(u_0, \ldots, u_{j-1}, A_0, A_1, A_2, \ldots\right)$$

$$P(j) = \text{diag}\left(\psi^{-j-q}\right)K_d'\left[u_0\psi^P, x\right]\text{diag}\left(\psi^{j+p}\right)\Big|_* ,$$

where * means that the expression is to be evaluated at grad $u_0 = 0$, grad$\chi = A_0$. When determinant of $P(j)$ vanishes for certain values of j, we get the resonance positions. The number N of these is usually equal to the order of the system defined by the dominant part $K_d[u, x]$. Whenever j reaches a positive integer index i, the system $E_j = 0$ is singular. In such a case, its general solution is the sum of a particular solution of the inhomogeneous system, and of the general solution of the homogeneous system whose dimension is that of kernel of $P(i)$.

To arrive at a representation of the general solution without introducing movable logarithmic terms, it is necessary that the dimensionality of this kernel be exactly the multiplicity of i, hence the two necessary conditions to be satisfied at any positive integer index i:

1. multiplicity of i to be equal to the dim Ker $P(i)$ (3.7.10)

2. vector Q_i orthogonal to the kernel of the adjoint of $P(i)$

We can now summarize the necessary conditions to be satisfied for the Painleve property of a nonlinear system to be

1. For every branch (p, u_0), all components p are integers.

2. All resonances are integers.

3. For any index i, the condition (3.7.10) is satisfied.

4.   For any index i, and any perturbation order $n \geq 0$, the vector Q_i^n is orthogonal to the kernel of the adjoint operator $P(i)$.

The above discussion can be better understood from an elegant example discussed in reference (61).

Example:

Consider the equation

$$K[u] = u_{xx} + 3uu_x + u^3 = 0 \ ,$$

which is nothing but the stationary form of the second member of Burger's hierarchy. One can easily verify that

$$u = 1/(x-a) + 1/(x-b)$$

is a general solution. The usual Painleve expansion yields

(1) Singularity of order $p = -1$, $u_0^0 = 1$ resonances $(-1, 1)$.

(2) Singularity of order $p = -1$, $u_0^0 = 2$ resonances $(-2, -1)$.

We have chosen the gauge $S = 0$. In the first case, one gets

$$u^0 = \psi^{-1} + c_1 - c_1^2 \psi + c_1^3 \psi^2 - c_1^4 \psi^3 + \ldots \ ,$$

which is nothing but

$$u^0 = \psi^{-1} + \frac{c_1}{1 + c_1 \psi} \ ; \quad 0 < |\psi| < |c_1|^{-1} \ ,$$

whereas for (2) the Painleve series reduces to single term

$$u^0 = 2\psi^{-1} \ .$$

If we now perturb these solutions as

$$u = \sum_{n=0}^{\infty} \varepsilon^n u^n ; \quad K[u] = 0 ; \quad K\left[u^0\right] = 0 ,$$

then the linearized equation is defined by the operator

$$K'[u] = \partial_x^2 + 3u\partial_x + 3u_x + 3u^2 .$$

The first few terms of the expansion are determined from

$$u^{(1)} : K'\left[u^0\right]u^1 = 0$$

$$u^2 : K'\left[u^0\right]u^2 = -3u^{(1)}u_x^1 - 3u^0\left(u^1\right)^2$$

$$u^3 : K'\left[u^0\right]u^3 = -3\left(u^1 u^2\right)_x - \left(u^1\right)^3 .$$

One should now use the particular solutions of $K'\left[u^0\right]\omega = \psi^{N-2}$ defined by

$$u^0 = \chi^{-1} : \omega = \frac{1}{N(N+2)}\psi^N , \quad N(N+2) \neq 0$$

$$u^0 = 2\chi^{-1} : \omega = \frac{1}{(N+2)(N+3)}\psi^N , \quad N \neq -2, -3 .$$

One can calculate the desired expansions, which in this simple case can be written in closed form:

$$u = \left(\varepsilon A_1 \sum_{j=-\infty}^{-2} (\varepsilon A_1)^{-j-2} \psi^j \right) + \psi^{-1} + \left(C_1 \sum_{j=0}^{\infty} (-C_1\psi)^j \right)$$

$$= \frac{\varepsilon A_1 \psi^{-2}}{1 - \varepsilon A_1 \psi^{-1}} + \psi^{-1} \frac{C_1}{1 + C_1\psi} = \frac{1}{\chi - \varepsilon A_1} + \frac{C_1}{1 + C_1\chi} .$$

$C_1 = C_1 + \varepsilon B_1, ..., A_1, B_1$ are arbitrary coefficients, so the net effect of the perturbation is to shift the two movable PDEs from one arbitrary location to the other. If one starts with $C_1 = 0,$ then the perturbation will shift the single pole at $\chi = 0$ to the nearby point $\chi = \varepsilon A_1$, and the new pole $\chi = -1/\varepsilon B_1$ to ∞. Similarly, for the second family one obtains

$$u = 2\chi^{-1} + \varepsilon\left(A_1\chi^{-3} + B_1\chi^{-2}\right) + \varepsilon^2\left(\frac{1}{2}A_1^2\chi^{-5} + \frac{3}{2}A_1B_1\chi^{-4}\right) + \varepsilon^2\left(\frac{1}{4}A_1^3\chi^{-7} + \ldots\right) .$$

Identifying this with the solution

$$u = \sum_{j=0}^{\infty}\left(a^j + b^j\right)\chi^{-j-1} ,$$

we find

$$a + b = \varepsilon B_1, \quad a^2 + b^2 = \varepsilon A_1$$

and

$$u = 2\chi - \varepsilon B_1 \Big/ \left\{\chi^2 - \varepsilon B_1\chi + 1/2\left(-\varepsilon A_1 + \varepsilon^2 B_1^2\right)\right\} , \tag{3.7.10}$$

where A_1, B_1 are arbitrary constants at level one. The simple pole at $\chi = 0$ with residue 2 has been unfolded into two simple poles with residue at locations $1/2\left(\varepsilon B_1 \pm \sqrt{2\varepsilon A_1 - \varepsilon^2 B_1^2}\right)$ both very near to 0. This example illustrates the fact that both the principal family and the nonprincipal family give rise to a local representation of the general solution. Various other complicated situations have been discussed in the paper of Conte *et al.*, and the interested reader is referred to those for detailed information.

3.8 Painleve analysis of perturbed equation

While the above approach tries to explain the relevance of negative resonances in the Painleve analysis of a particular nonlinear equation, another method was adopted by Basak and Roy Chowedhury[97] for studying the Painleve properties of equations that are nearly integrable, that is perturbation of some exactly integrable systems such as the nonlinear Schrodinger equation.

The perturbed NLS equation is written as

$$q_t = iq_{xx} + 2i|q|^2 q + \left(a_1 q_{xxx} + a_3 q^2 q_x\right)$$
$$q_t = -iq_{xx}^* - 2i|q|^2 q^* + \varepsilon\left(a_1 q_{xxx}^* + a_3 q^{*2} q_x\right) . \tag{3.8.1}$$

The leading order exponent is found to be equal to -1, so we substitute

$$q = \sum_{j=0}^{\infty} u_j \phi^{j-1}, \quad q^* = \sum_{j=0}^{\infty} v_j \phi^{j-1}.$$

The recursion relation for the coefficients are found to be

$$u_{j-3,t} + (j-3)u_{j-2}$$

$$= j(j-2)(j-3)\phi_x^2 U_{j-1} + j(j-3)\left[2\phi_x u_{j-2,x} + \phi_{xx} u_{j-2} + i u_{j-3xx}\right.$$

$$+ \sum_m \sum_n \{2i u_m u_n v_{j-m-n-1}\} + \varepsilon a_1 \left[(j-1)(j-2)(j-3)\phi_x^3 u_j\right.$$

$$+ (j-2)(j-3)\left(3u_{j-1x}\phi_x^2 + 3u_{j-1}\phi_x\phi_{xx}\right) + (j-3)\left(3u_{j-2xx}\phi_x\right.$$

$$+ 3u_{j-2x}\phi_{xx} + u_{j-2}\phi_{xxx}\right) + u_{j-3xxx}\right]$$

$$+ \varepsilon a_3 \left[\phi_x \sum_m \sum_n u_m u_n v_{j-m-n}(j-m-n-1) + \sum_m \sum_n u_m u_{nx} v_{j-m-n-1,x}\right]\right],$$

along with

$$v_{j-3t} + (j-3)v_{j-2}\phi_t$$

$$= -j(j-2)(j-3)\phi_x^2 v_{j-1} - j(j-3)\left(2\phi_x v_{j-2x} + \phi_{xx} v_{j-2}\right)$$

$$- i v_{j-3xx} - 2i \sum_m \sum_n v_m v_n u_{j-m-n-1} + \varepsilon a_1\left[(j-1)(j-2)(j-3)\phi_x^3 v_j\right.$$

$$+ (j-2)(j-3)\left(3\phi_x^2 v_{j-1x} + 3\phi_x\phi_{xx} v_{j-1}\right) + (j-3)\left(3v_{j-2xx}\phi_x\right.$$

$$+ 3v_{j-2x}\phi_{xx} + v_{j-2}\phi_{xxx}\right) + v_{j-3xxx}\right] + \varepsilon a_3\left[\phi_x \sum_m \sum_n v_m v_n\right.$$

$$\times u_{j-m-n}(j-m-n-1) + \sum_m \sum_n v_m v_n u_{j-m-m-1x}\right]. \tag{3.8.2}$$

Setting $j = 0$, we get

$$u_0 v_0 = -6a_1/a_3 \cdot \phi_x^2. \tag{3.8.3}$$

Collecting coefficients of u_j and v_j,

$$\varepsilon a_1\left[(j-1)(j-2)(j-3)\phi_x^3\right]u_j + \varepsilon a_3\left[\phi_x u_0^2 v_j(j-1) - 2u_0 v_0 \phi_x u_j\right] = \text{terms}$$

containing u_{j-1}, v_{j-1} etc.

$$\varepsilon a_1\left[(j-1)(j-2)(j-3)\phi_x^3\right]v_j + \varepsilon a_3\left[\phi_x v_0^2 u_j(j-1) - 2u_0 v_0 \phi_x v_j\right] = \text{terms}$$

containing u_{j-1}, v_{j-1} etc.

So, we get that the resonances are determined from the conditions

$$\det\begin{vmatrix} a_1(j-1)(j-2)(j-3)\phi_x^3 - 2a_3 u_0 v_0 \phi_x & a_3\phi_x u_0^2(j-1) \\ a_3\phi_x v_0^2(j-1) & a_1(j-1)(j-2)(j-3)\phi_x^3 - 2a_3 u_0 v_0 \phi_x \end{vmatrix} = 0.$$

(3.8.3)

Using equation (3.8.3), we get

$$j(j+1)(j-3)(j-4)\left(j^2 - 6j + 17\right) = 0 ,$$

so we get resonances at

$$j = 0, -1, 3, 4$$

and at the roots of $j^2 - 6j + 17 = 0$.

Unfortunately, the roots of this equation are not integers. But remember that the order of the equation is four. Now for $j = 1$, equation (3.8.2) yields m,

$$2i\phi_x^2 u_0\left(1 - \frac{6a_1}{a_3}\right) + 6\varepsilon a_1 \phi_x(u_{0x}\phi_x + u_0\phi_{xx})$$

$$+ \varepsilon a_3\left(u_0^2 v_{0x} - 2\phi_x u_0 v_0 u_1\right) = 1$$

(3.8.4)

$$-2i\phi_x^2 v_0\left(1 - \frac{6a_1}{a^3}\right) + 6\varepsilon a_1 \phi_x(v_{0x}\phi_x + v_0\phi_{xy})$$

$$+ \varepsilon a_3\left(v_0^2 u_{0x} - 2\phi_x u_0 v_0 v_1\right) = 0 .$$

(3.8.5)

Multiplying (3.8.5) by u_0 and (3.8.4) by v_0, we add and subtract, then we obtain

$$4i\phi_x^2 u_0 v_0 \left(1 - \frac{6a_1}{a_3}\right) + 6\varepsilon a_1 \phi_x^2 (u_{0x} v_0 - u_0 v_{0x})$$

$$+ \varepsilon a_3 v_0 u_0 \left[-(v_{0x} v_0 - v_{0x} u_0) + 2\phi_x (v_0 v_1 - v_0 u_1)\right] = 0 ,$$ (3.8.6)

along with

$$u_0 v_1 + v_0 u_1 = \frac{6a_1}{a_3} \phi_{xx} .$$ (3.8.7)

Now equation (3.8.6) can be of order ε only if

$$1 - \frac{6a_1}{a_3} = 0 ,$$

whence we get from (3.8.7)

$$u_0 v_1 + v_0 u_1 = \phi_{xx}$$

$$v_1 u_0 - v_0 u_1 = (u_{0x} v_0 - v_{0x} u_0)/\phi_x$$

$$u_0 v_0 = -\phi_x^2 .$$

Now setting $j = 2$ in the recursion relation we obtain

$$-u_0 \phi_t = -i\left(2u_{0x}\phi_x + u_0\phi_{xx}\right) + 2iu_0^2 v_1 + 4iu_0 v_0 u_1 + \varepsilon a$$

$$+ \varepsilon a_1 \left[-(3u_{0xx}\phi_x + 3u_{0x}\phi_{xx} + u_0\phi_{xxx})\right] + \varepsilon a_3 \left[\left(\phi_x u_0^2 v_2\right.\right.$$

$$\left.\left. -\phi_x u_1^2 v_0 + u_0^2 v_{1x} + 2u_0 u_1 v_{0x}\right)\right] - v_0 \phi_t .$$ (3.8.8)

Since we want to truncate the series at a constant level, we set $u_j, v_j = 0,\ j > 2$ so that

$$u_{0x} v_0 - u_0 v_{0x} = \lambda \phi_x^2$$

$$2\phi_t/\phi_x = \varepsilon a_1 \left(2\psi_x - \psi^2 + \frac{3}{2}\lambda^2\right) ,$$ (3.8.9)

where

$$\psi = \phi_{xx}/\phi_x \,, \quad u_1 v_1 = \frac{1}{4}\left(\lambda^2 - \psi^2\right) .$$

On the other hand, for $j = 3$ we have

$$u_{0t} = iu_{0xx} + 2iu_1^2 v_0 + 4iu_0 u_1 v_1 + \varepsilon q_1 u_{0xxx}$$

$$v_{0t} = -iv_{0xx} - 2iv_1^2 u_0 - 4iv_0 v_1 u_1 + \varepsilon a_1 v_{0xxx} \,,$$

from which we deduce that

$$2\frac{\phi_t}{\phi_x} = \varepsilon a_1 \left[2\psi_{xx} - \psi^3 + \frac{3}{2}\lambda^2 \psi\right] .$$

It is interesting to observe that this equation is nothing but a total derivative of equation (3.8.9),

$$\frac{\partial}{\partial x}\left\{\frac{2\phi_t}{\phi_x} - \varepsilon a_1\left(2\psi_x - \psi^2 + \frac{3}{8}\lambda^2\right)\right\} = 0 \,, \tag{3.8.10}$$

so they are compatible.

On the other hand, equations (3.8.2) and (3.8.5) lead to

$$\frac{v_0^2}{\phi_x^2}\frac{\partial}{\partial t}\left(\frac{u_0}{v_0}\right) = i\left(\psi^2 - 2\psi_x - \frac{3}{2}\lambda^2\right) + \varepsilon a_1 \lambda\left(2\psi^2 - 4\psi_x + \frac{\lambda^2}{4}\right) .$$

A similar consideration at $j = 4$ gives rise to

$$\left(u_1 v_1\right)_t = i\frac{\partial}{\partial x}\left(v_1 u_{1x} - u_1 v_{1x}\right) + \varepsilon a_1\left[\left(u_1 v_1\right)_{xxx} - 3\left(u_{1x} v_{1x}\right)_x\right] + 6\varepsilon a_1\left(u_1 v_1\right)_x u_1 v_1$$

$$v_1^2 \frac{\partial}{\partial t}\left(\frac{u_1}{v_1}\right) = i\left[\left(u_1 v_1\right)_{xx} - 2u_{1x} v_{1x}\right] + 4i\left(u_1 v_1\right)^2$$

$$+ \varepsilon a_1\left(v_1 u_{1xxx} - u_1 v_{1xxx}\right) - 6\varepsilon a_1 u_1 v_1\left(v_1 u_{1x} - u_1 v_{1x}\right) .$$

Putting the values of u, v, we get compatibility with (3.8.10)

$$\lambda\left[\left\{\phi_j x\right\}_x \big/ \psi + \psi^2 \big/ 4 + 1 / 2\{\phi, x\}\right] = 0 , \qquad \lambda^2 \{\phi, x\}_x = 0 ,$$

where $\{\phi, x\}$ is the Schwarzian derivative.

λ is a constant of integration. If we choose $\lambda = 0$, we at once get

$$i\left(\psi^2 - 2\psi_x\right) = -i / \psi\left(2\psi_{xx} - \psi^2\right) ,$$

whose consequence is

$$\psi_{xx} - \psi\psi_x = 0 \quad \text{or} \quad \{\phi, x\}_x = 0 .$$

The singular manifold ϕ is no longer arbitrary. Therefore, one can conclude that the perturbed NLS equation does not pass the Painleve test in totality, but it can have partial integrability. It may be added that the perturbed Painleve analysis of Conte *et al.* and the Painleve analysis of perturbed systems are totally different ideas.

3.9 Painleve equation and Monodromy matrix

As per the ARS conjecture, the equations that are integrable in the sense that they have Lax pairs will reduce to various nODEs in some scaled variables. These nODEs also inherit a different sort of Lax pair from the original one, and its property can be analyzed via these linear sets. On the other hand, in many cases these reduced ordinary equations belong to the Painleve class, so it is interesting to study the properties of the ordinary Painleve system via Lax pair.

Monodromy matrix and Stoke's parameters are two of the most important ingredients in such a study. The initial rigorous study of the monodromy property of a linear differential equation was done by the Kyoto School mainly led by Ueno and Date[102] and Jimbo and Miwa.[81] The other approach was due to Flaschka and Newell.[103] Although the former tries to set up a general abstract formalism, the latter is more concrete and oriented to a special problem at hand.

The monodromy approach can be used to set up an alternative approach to inverse scattering transform. While the inverse scattering method tries to reconstruct the potential from the Scattering data, the monodromy approach tries to do the same from the information of the nature of singularities of the linear problem and the associated Stoke's parameter. The linerature of such analysis is quite old, and one should mention the classic analysis by Sibuya, which gives explicit prescription for the determination of the Stoke's parameter. A well-known application of monodromy matrix can be found in the theory of hypergeometric functions, where they are used to connect solutions in the neighborhood of one singularity with that in the neighborhood of another one. In this section we proceed to show how analysis of monodromy matrix and Stoke's parameter proceeds in case of nODEs obtained as a reduction of a nPDE, which is integrable. The following details two examples related to the nonlinear Schrodinger equation and three wave interaction problems. The former one reduces to a coupled set equivalent to Painleve IV, and the latter yields an ordinary equation that is not a Painleve family but can be connected to Painleve VI via a complicated transformation.

Example 1: NLS problem[70]
The nonlinear Schrodinger equation can be written as

$$iq_t - q_{xx} = \pm 2q^2 q^* . \tag{3.9.1}$$

The AKNS inverse problem pertaining to (3.9.1) is

$$v_{1x} = -i\bar{\xi}v_1 + qv_2$$
$$v_{2x} = i\bar{\xi}v_2 + rv_1 , \tag{3.9.2}$$

along with

$$v_{1t} = Av_1 + Bv_2$$
$$v_{2t} = Cv_1 + Dv_2 , \tag{3.9.3}$$

with the form of A, B, C, D, well known as functions of q, r, $\bar{\xi}$. Lie symmetry suggests that a similarity variable is given by

$$z = xt^{-1/2} , \quad q(x,t) = t^{1/2}\phi\left(x/t^{1/2}\right)$$

when equation (3.9.1) is changed to

$$-d/dz(\phi + iz/2 \cdot \phi) = \pm 2\phi^2\phi^*$$

and its conjugate.

The main trick of Flaschka and Newell is to transform the Lax pair to such variables, for which we set

$$v = \begin{pmatrix} v_1 \\ v_2 \end{pmatrix} \qquad \begin{aligned} v_i &= v_i\left(xt^{-1/2}, \ \bar{\xi}t^{1/2}\right) \\ &= v_i(z, \xi) \end{aligned}$$

so that one gets

$$\begin{aligned} v_z^1 &= -i\xi v'^1 + \phi v^2 \\ v_z^2 &= i\xi v^2 + \phi^* v^1 \end{aligned}$$

(3.9.4a)

and

$$v_\xi^1 = \left(4i\xi + \frac{2i\phi\phi^*}{\xi} - iz \right)v^1 + \left(-4\phi - \frac{2i\phi_z}{\xi} + \frac{2\phi}{\xi} \right)v^2$$

$$v_\xi^2 = \left(-4\phi^* + \frac{2i\phi_z}{\xi} + \frac{2\phi}{\xi} \right)v' + \left(-4i\xi - \frac{2i\phi\phi^*}{\xi} + iz \right)v^2 .$$

(3.9.4b)

The last set can be written compactly as

$$v_\xi = \left[A_0\xi + A_1 + \frac{1}{\xi} \cdot A_2 \right]v$$

(3.9.5)

$$A_1 = \begin{pmatrix} 4i & 0 \\ 0 & -4i \end{pmatrix} ; \qquad A_1 = \begin{pmatrix} -iz & -4\phi \\ -4\phi^* & iz \end{pmatrix}$$

$$A_2 = 2i \begin{bmatrix} \phi\phi^* & -\left\{\phi_z + \dfrac{i}{2}z\phi\right\} \\ \left\{\phi_z^* - \dfrac{i}{2}z\phi^*\right\} & -\phi\phi^* \end{bmatrix} .$$

Asymptotic expansion

The analysis of the Lax pair starts with an asymptotic expansion in ξ. To proceed, we set

$$v_\chi - \exp\left(a_0\xi^2 + a_1\xi\right)\xi^\mu \sum_k C_k\xi^{-k} ,$$

(3.9.6)

whence equation (3.9.5) leads to

$$[2a_0\xi + a_1 + (\mu - k/\xi)]\exp(a_0\xi^2 + a_1\xi)\xi^u \sum C_k\xi^{-k}$$
$$= \left(A_0\xi + A_1 + A_2\xi^{-1}\right)\exp(a_0\xi^2 + a_1\xi)\xi^u \sum C_k\xi^{-k} \ .$$

Comparing the coefficients of various powers of ξ, we can find two independent sets of solution:

$$\bar{v}_\infty(1,z,\xi) \approx \exp\left(2i\xi^2 - iz\xi\right)\left[\begin{pmatrix} 1 \\ 0 \end{pmatrix} + \xi^{-1}\begin{pmatrix} Q \\ \frac{i}{2}\phi^* \end{pmatrix} + \ldots\right]$$

$$\tilde{v}_\infty(2,z,\xi) \approx \exp\left(-2i\xi^2 + iz\xi\right)\left[\begin{pmatrix} 0 \\ 1 \end{pmatrix} + \xi^{-1}\begin{pmatrix} -i/2\phi \\ Q \end{pmatrix} + \ldots\right] \ .$$

Now one may note that equation (3.9.4) implies

$$Q = \phi\phi_z^* - \phi^*\phi_z - \frac{1}{2}z\phi\phi^* = \frac{i}{2}\int \phi\phi^* dz + C \ .$$

The above equation gives us the behavior near $\xi = \infty$. To obtain the expansion near $\xi = 0$, we put $\xi = 1/\eta$ and consider the behavior of v near $\eta \to \infty$. Under such a transformation, equation (3.9.5) is transformed to

$$v_\eta = -\left(A_2\eta^{-1} + A_1\eta^{-2} + A_0\eta^{-3}\right)v \ .$$

Setting $v = \eta^u \sum C_k\eta^{-k}$, we can proceed as before and obtain the following expansion near $\xi = 0$:

$$\tilde{v}_0(1,z,\xi) = e^{v(z)}\xi^{-2k}\left[\begin{pmatrix} R \\ 1 \end{pmatrix} + \frac{\xi^{-1}}{1-4k}\begin{pmatrix} x_2 \\ y_2 \end{pmatrix} + \ldots\right] ,$$

where

$$R = \frac{i\phi_z + \frac{i}{2}z\phi}{i\phi\phi^* + k}$$

and

$$\begin{pmatrix} x_2 \\ y_2 \end{pmatrix} = \begin{pmatrix} R^* \\ 1 \end{pmatrix} S \qquad (3.9.7)$$

$$S = -2i\left(\phi_z^* - \frac{i}{2}z\phi^*\right)(4\phi - R) - (2i\phi\phi^* + 2k - 1)(iz - 4\phi^* \cdot R) \ .$$

The second kind of solution turns out to be

$$\tilde{v}_0(2, z, \xi) = e^{u^*(z)}\xi^{2k}\left[\begin{pmatrix} 1 \\ R^* \end{pmatrix} + \ldots\right] , \qquad (3.9.8)$$

where

$$u = \int \phi \, dz \ .$$

The condition for degeneracy for C_0 yields

$$\det[A_2 + Iu] = 0 \ ,$$

which can be written as

$$\mu^2 = A\left[\left(\phi_z + \frac{i}{2}z\phi\right)\left(\phi_z^* - \frac{1}{2}iz\phi^*\right) - (\phi\phi^*)^2\right] = 2k \quad \text{and} \quad \frac{d\mu^2}{dz} = 0 \ .$$

At this point, we mention some important features of equations (3.9.4a) and (3.9.4b):

(a) If $v(1, \xi, z)$ is a solution, then

$$Mv^*(1, \xi^*, z)$$

is also a solution.

(b) If $Mv^*(2, \xi^*, z)$ is a solution, then

$$\tilde{v}(2, \xi, z)$$

is another solution.

Here $M = \begin{pmatrix} 0 & 1 \\ 1 & 0 \end{pmatrix}$, $v(n, \xi, z)$ denotes the solution vector with $n = 1, 2$, indicating the first and second kind solution.

Growth and decay of solution

It is important to remember that the study of growth and decay of the solution vector yields information about the Stoke's parameter. Here we try to segregate the zone in the complex plane where the solutions defined in the above section show definite pattern of dominancy or subdominancy. In the following figure we have depicted this division of the complex eigenvalue plane into several sectors. Recall that the lines in the ξ-plane originating from the origin serves this purpose, depending on the behavior of v. In our present situation, $\arg \xi = \pi / 4$, $3\pi / 4$, $5 / 4\pi$, $7 / 4\pi$ are Stoke's lines, and the sectors are defined as

$$S_j = \xi, \quad |\xi| > \rho, \quad \text{for some } \rho,$$

with

$$1 / 2(j-1)\pi \le \arg \xi < \frac{1}{2} j\pi, \quad j = 1, 2, \dots .$$

The anti-Stoke's lines are

$$\arg \xi = 0, \quad \pi / 2, \quad \pi, \quad 3\pi / 2, \quad 2\pi .$$

For the ensuing discussion, we fix a new notation for the solution v.

$v_j^{(1)}(k, \xi, z)$ will denote a solution of the linear equation (4a,b), where "i" denotes first and second component as above, j denotes sector, and k denotes the type of solution. The next important stage is to write down the basic form of the matrices connecting these solution vectors in the several sectors. One should note that a solution that was dominating in one sector may become subdominant when its leading order terms are cancelled by the contribution from other components in the other sector. This fact implies that the connection matrices are all triangular. We have

$$v_2 = v_1 \begin{pmatrix} 1 & 0 \\ a & 1 \end{pmatrix}; \quad v_3 = v_2 \begin{pmatrix} 1 & b \\ 0 & 1 \end{pmatrix};$$

$$v_4 = v_3 \begin{pmatrix} 1 & 0 \\ c & 1 \end{pmatrix}; \quad v_5 = v_4 \begin{pmatrix} 1 & d \\ 0 & 1 \end{pmatrix}; \quad \text{but } v_5 = v_1 .$$

$$(3.9.9)$$

Using equation (3.9.9) and symmetry properties noted above, we obtain

$$v_2^{(1)} = v_1^{(1)} + av_1^{(2)} ; \quad v_4^{(1)} = v_3^{(1)} + cv_3^{(2)} \tag{3.9.10}$$

and so

$$c = a , \quad d = b .$$

Also, for $\pi \le \arg \xi < \frac{3}{2}\pi$,

$$v_3^{(1)} = bv_1^{(1)} + (1+ab)v_1^{(2)} . \tag{3.9.11}$$

For $\frac{3}{2}\pi \le \arg \xi < 2\pi$,

$$v_4(\xi, z) = v_3(\xi, z) \begin{pmatrix} 1 & 0 \\ c & 1 \end{pmatrix}, \tag{3.9.12}$$

from which it follows that

$$v_4^{(1)} = v_1^{(1)}(1+bc) + (a+c+abc)v_1^{(2)}$$
$$v_4^{(2)} = bv_1^{(1)} + (1+ab)v_1^{(2)} \tag{3.9.13}$$

Crossing this zone, we come back to the first sector again, hence

$$v_5^{(1)} = v_4^{(1)} = (1+bc)v_1^{(1)} + (a+c+abc)v_1^{(2)}$$
$$v_5^{(2)} = dv_4^{(1)} + v_4^{(2)} = v_1^{(1)}[d(1+bc)+d] + [1+ad+d(a+c+abc)]v_1^{(2)} \tag{3.9.14}$$

These relations will be useful when we connect the solution near the origin to that in the neighborhood of ∞. From equation (3.9.7), we observe that

$$\overline{v}(1, \xi, z) = v(1, \xi, z) - jiv(2, \xi, z) \ln \xi , \tag{3.9.15}$$

where j stands for

$$j = \left[2\left(\phi_2^* - \frac{1}{2}z\phi^*\right)(\phi - R) - (2i\phi\phi^* + 2k - 1)(iz - 4i\phi^* R)\right] \times \exp[u - u^*] .$$

The logarithm will disappear if $j = 0$, $k = 1/4$. Furthermore, from (3.9.15), we get

$$M\bar{v}\left(1,\xi,e^{-i\pi},z\right) = e^{2i\pi k}\bar{v}(1,\xi,z) - \pi j e^{-2\pi i k}\bar{v}(2,\xi,z)$$

$$M\bar{v}\left(2,\xi e^{-i\pi},z\right) = e^{-2i\pi k}v(2,\xi,z) \ .$$

So, if in the sector $0 \le \arg < 2\pi$ the solution is $\bar{v}$ then $\bar{v}\left(\xi e^{2\pi i},v\right) = v(\xi,z)J$, a fundamental solution in $(2\pi,\ 4\pi)$ where

$$J = \begin{pmatrix} e^{-4i\pi k} & 0 \\ 2\pi j e^{4i\pi k} & e^{4\pi i k} \end{pmatrix} \ .$$

Note that $\det J = 1$ for all K. Now we seek the matrix connecting v_0 and v_∞ as

$$v_\infty = v_0 A \ , \tag{3.9.16}$$

where

$$A = \begin{pmatrix} \alpha & \beta \\ \gamma & \delta \end{pmatrix} \ .$$

Now,

$$\det v_\infty = 1$$

and

$$\det v_0 = \frac{-2ik\left(\phi_z + \dfrac{i}{2}z\phi\right)}{(i\phi\phi^* + k)(i\phi\phi^* - k)} \ .$$

Hence,

$$\det A = \frac{(i\phi\phi^* + k)(i\phi\phi^* - k)}{-2ik\left(\phi_z + \dfrac{i}{2}z\phi\right)} \ .$$

The set $(a,b,c,d,\alpha,\beta,\gamma,d,\det v_\infty)$ and the coefficients of the asymptotic expansions are called the monodromy data.

Properties of A:

From equation (3.9.16) it follows that

$$v_\infty\left(\xi e^{2\pi i}\right) = v_0\left(\xi e^{2\pi i}\right)A$$
$$= v_0(\xi)JA \ ,$$

and in the last sector

$$v_5\left(\xi e^{2\pi i}\right) = \begin{pmatrix} 1 & 0 \\ a & 1 \end{pmatrix}\begin{pmatrix} 1 & b \\ 0 & 1 \end{pmatrix}\begin{pmatrix} 1 & 0 \\ a & 1 \end{pmatrix}\begin{pmatrix} 1 & b \\ 0 & 1 \end{pmatrix}v_1\left(\xi e^{2\pi i}\right) \ ,$$

leading to

$$\begin{pmatrix} 1 & 0 \\ a & 1 \end{pmatrix}\begin{pmatrix} 1 & b \\ 0 & 1 \end{pmatrix}\begin{pmatrix} 1 & 0 \\ a & 1 \end{pmatrix}\begin{pmatrix} 1 & b \\ 0 & 1 \end{pmatrix} = A^{-1}J^{-1}A \ .$$

We now set $\xi = \hat{\xi}\xi^{-i\pi}$ in the solution v_∞ and apply $M = \begin{pmatrix} 0 & 1 \\ 1 & 0 \end{pmatrix}$, whence we get the following important and fundamental relation:

$$Mv^{(1)} = \left(\alpha\delta e^{2\pi ik} + \alpha\beta\pi je^{-2\pi ik} - \beta\gamma e^{-2\pi ik}\right)v^{(1)}$$
$$+\left[-\alpha\gamma\left(e^{2\pi ik} - e^{-2\pi ik}\right) - \alpha^2\pi je^{-2\pi ik}\right]v^{(2)}$$

$$Mv^{(2)} = \left[\beta\delta\left(e^{2\pi ik} - e^{-2\pi ik} + \beta^2\pi je^{-2\pi ik}\right)\right]v^{(1)}$$
$$+\left(-\beta\gamma e^{2\pi ik} - \alpha\beta\pi je^{-2\pi ik} + \delta\alpha e^{-2\pi ik}\right)v^{(2)} \ .$$

From this, one can deduce that

$$l = \alpha\delta e^{-\pi ik} + \alpha\beta\pi je^{-2\pi ik} - \beta\gamma e^{-2\pi ik}$$

$$a = -2\pi\gamma\sin(2\pi k) - \pi j\alpha^2 e^{-2\pi ik}$$

$$b = \sin(2\pi k) + \pi j\beta^2 e^{-2\pi ik}$$

$$l + ab = -\beta\gamma e^{2\pi ik} - \alpha\beta\pi je^{-2\pi ik} + \alpha\delta e^{-2\pi ik} \ .$$

Sibuya's approach to Stoke's parameters

One of the earliest work on the monodrony theory and the Stoke's parameter was done by Sibuya. He introduced an algorithm by which it is possible to have an approximate (sometimes exact) determination of the Stoke's parameter. In reference, Sibuya treats the equation

$$\frac{d^2 v}{d\xi^2} - \left(\xi^\mu + a_1 \xi^{\mu-1} + \ldots + a_{\mu-1}\xi + a_\mu \right) v = 0 \tag{3.9.17}$$

under the following assumptions:

(i) The differential equation (3.9.17) has a unique solution.

(ii) v is an entire function of its parameters

$$v = v_\mu\left(\xi, a_1, \ldots, a_\mu \right) .$$

(iii) v admits an asymptotic representation

$$v \sim \xi^{\alpha_\mu}\left(1 + \sum_{n=1}^{\infty} B_{\mu,n}\xi^{-n/2} \right) \exp\left[-iE_\mu(\chi, t) \right] ,$$

as χ tends to infinity in the different sectors, where $E_\mu(x,t)$ are represented as

$$E_\mu(x,t) \approx \frac{2}{\mu+2}\xi^{(u+2)/2} + \sum_{n=1}^{\mu+1} A_{\mu,n}\xi^{(u+2-n)/2}$$

where α_μ, $A_{\mu,n}$, $B_{\mu,n}$ are polynomials in $\left(a_1, \ldots, a_\mu \right)$. We now collect some of the results of Sibuya in the following:

(a) Suppose we set

$$\left(1 + a_1\xi^{-1} + \ldots + a_\mu\xi^{-\mu} \right)^{1/2} + 1 + \sum_{k=1}^{\infty} b_k\xi^{-k} ,$$

then the quantities γ_u and $A_{u,n}$ are given by

$$\gamma\mu = \begin{cases} -\mu/4 & \dots \mu \text{ odd} \\ -\mu/4 - b_{u/2} + 1 \dots \mu \text{ even} \end{cases}$$

and

$$\sum_{n=1}^{\mu+1} A_{\mu,n} x^{(\mu+2-n)/2} = \sum_{1 \le n \le \mu/2+1} \frac{2}{\mu+2-2n} b n x^{(\mu+2-n)/} \ .$$

Next choose ϕ such that

$$\exp[i(\mu+2)\phi] = 1 \ ,$$

then the function $v\left(\xi, e^{i\phi} a_1, \dots, e^{i\phi\mu a\mu}\right)$ is also a solution. With $\theta = \exp\left(i\left(\frac{2\pi}{\mu} + 2\right)\right)$, the solution in the jth sector is given by

$$v_j(\xi,t) \sim \theta^{-jr_{\mu-1}} x'^{\mu j}\left(1 + \sum_{n=1}^{\infty} b_{\mu,n_j} x^{-n/2}\right) \exp[\Sigma] \ ,$$

where

$$\Sigma = (-)^{j+1} i E_\mu(\xi,\alpha) \ ,$$

as $\xi \to \infty$ in the sectors.

(iv) Let us now consider two solutions, $v_{\mu j+1}$ and $v_{\mu j+1}$, which are linearly independent because $v_{\mu j+1}$ is subdominant in the $(j+1)$th sector and $v_{\mu j+2}$ is dominant. Therefore, v_μ is a combination of $v_{\mu j+1}$ and $v_{\mu j+2}$;

$$v_j(\chi,t) = C_j(t) v_{j+1} + \tilde{C}_j(t) v_{j+2} \ ,$$

where C_j, $\tilde{C}_j$ are Stoke's multipliers. For $\mu = 2$,

$$C_j(a_1,a_2) = \begin{cases} 2^{b_2} \exp\left[\frac{1}{4}a_1^2 - i\pi\left(\frac{b_2}{2} - \frac{1}{4}\right)\right]\dfrac{\sqrt{2\pi}}{\Gamma(1/2+b_2)} & ; \quad j \text{ even} \\[4mm] 2^b - \exp\left[-\frac{1}{4}a_1^2 + i\pi\left(\frac{b_2}{2} + \frac{1}{4}\right)\right]\dfrac{\sqrt{2\pi}}{\Gamma(1/2-b_2)} & ; \quad j \text{ odd} \end{cases}$$

$$\tilde{C}_j = \begin{cases} -i\exp(-\pi b_2) & j \text{ even} \\ -i\exp(\pi b_2) & j \text{ odd} \end{cases}$$

$$b_2 = \frac{1}{2}a_2 - \frac{1}{8}a_1^2 \ .$$

We now apply this result to our case of equation (3.9.4b). If we eliminate v_2, we obtain

$$v_{1\xi\xi} = \left\{ -16\xi^2 + 8z\xi - 4i - z^2 + \xi^{-1}\left[8i\left(\phi^*\phi_2 - \phi\phi_z^*\right) - 4z\phi\phi^*\right]\right.$$

$$\left. +\xi^{-2}\left[-2i\phi\phi^{*2} + \left(\phi_z + \frac{i}{2}z\phi\right)\left(\phi_z^* - \frac{1}{2}z\phi^*\right) + \dots \right. \right\} , \qquad (3.9.18)$$

where we have dropped higher terms in ξ^{-1} as we will be working near $\xi \to \infty$. Scaling the variables ξ, v as $2\xi = y$, $4v = v'$, we get

$$v'_{1yy} = \left[y^2 + zy - \left(i - \frac{z^2}{4}\right) - 4y^{-1}\left(\phi\phi_z^* - \phi^*\phi_z - \frac{i}{2}z\phi^*\phi\right) + \dots \right]v_1' \ .$$

We can now apply the result of Sibuya to this equation and obtain the information about the Stoke's parameters, though they will be approximate due to the nature of equation (3.9.18). To do that, we change from a vector to a matrix notation for the solution of equations (3.9.4a) and (3.9.4b). We write it as

$$v^{ij} = \begin{pmatrix} v^{11} & v^{12} \\ v^{21} & v^{22} \end{pmatrix} ,$$

where $\begin{pmatrix} v^{11} \\ v^{21} \end{pmatrix}$ is $\tilde{v}(1,z,\xi)$, and $\begin{pmatrix} v^{12} \\ v^{22} \end{pmatrix}$ stands for $\tilde{v}(2,z,\xi)$. We also attach a suffix "j" to denote the sector that is $v_{(j)}^{ij}$. Now we get

$$v_{(1)}^{11} = v_{(0)}^{11} + a v_{(0)}^{12} \qquad (3.9.19)$$

and from equation (3.9.10)

$$v_{(-1)} = C_{-1}v_{(0)} + \tilde{C}_{-1}v_{(1)}$$

or

$$v_{(1)} = \frac{1}{\tilde{C}_{-1}}v_{(-1)} - \frac{C_{-1}}{C_{-1}}v_{(0)} \ . \tag{3.9.20}$$

But we have the identification,

$$v_{(0)}^{11} = v_{(-1)} \ ; \quad v_1^{11} = v_{(1)} \ , \quad v_{(0)}^{12} = (-i\phi)v_{(0)} \ ,$$

which yields

$$a = \frac{1}{i\phi}C_{-1} \ , \quad \tilde{C}_{-1} = 1 \ .$$

Furthermore,

$$v_{(2)}^{12} = v_{(1)}^{12} + bv_{(1)}^{11} \ , \tag{3.9.21}$$

but

$$v_{(2)} = \frac{1}{i\phi}v_{(1)}^{12}$$

$$v_{(3)} = v_{(1)}^{11} \quad . \tag{3.9.22}$$

$$v_{(4)} = v_{(2)}^{12}$$

Comparing (3.9.21) and (3.9.22),

$$b = -i\phi C_2 \ , \quad \tilde{C}_2 = -1 \ . \tag{3.9.23}$$

Similarly, one can deduce

$$C = \frac{1}{i\phi}C_{-1} \quad d = -i\phi C_6$$
$$\tag{3.9.24}$$
$$\tilde{C}_{-1} = 1 \qquad \tilde{C}_6 = -1$$

with C_1 given by the expressions given in page 135 and $b_1 = -z/2$, $b_2 = 1/2\left(i - 1/2z^2\right)$.

Properties of the monodromy data

From the property that the matrix functions v_j are holomorphic in $S_j = \{\xi, |\xi| > 0\}$ and $1/2(j-1)\pi \le \arg \xi < 1/2\pi j$ and

$$v_j \sim \tilde{v}_j = \left(1 + \frac{C_1}{\xi} + \ldots\right)\begin{pmatrix} e^{\theta} & 0 \\ 0 & e^{-\theta} \end{pmatrix} \tag{3.9.25}$$

$$\theta = 2i\xi^2 - iz\xi \quad \text{as} \quad |\xi| \to \infty \quad \text{in } S_j$$

with $v_{j+1} = v_j A_j$, one can assert that there exists a matrix function w of the form

$$w(\xi) = \hat{W}(\xi)\begin{pmatrix} \xi^{-2k} & 0 \\ 0 & \xi^{2k} \end{pmatrix}. \tag{3.9.26}$$

$\hat{\omega}(\xi)$ is holomorphic and for $\xi \, \varepsilon \, S_1$, should have

$$v_1(\xi) = \omega(\xi)A \tag{3.9.27}$$

with

$$\det A = \frac{(i\phi\phi^* + A)(i\phi\phi^* - k)}{-2ik(\phi_z + 1/2z\phi)}. \tag{3.9.28}$$

Furthermore, the solution $v(\xi)$ has the symmetry

$$m\bar{v}^*(\xi)M = \bar{v}(\xi), \quad M = \begin{pmatrix} 0 & 1 \\ 1 & 0 \end{pmatrix}.$$

Also, the matrices A_j are independent of z. Differentiating $v_{j+1} = v_j A_j$ with respect to z and multiplying by v_{j+1}^{-1}, we obtain

$$v_{j+1z}v_{j+1}^{-1} = v_{iz}A_j v_{j+1}^{-1} = v_{ij}A_j\left(v_j A_j\right)^{-1} = v_{iz}v_j^{-1}.$$

$v_{ij}v_j^{-1}$ is well defined in a deleted neighborhood of $\xi = \infty$, and its asymptotic expansion is that of $\tilde{v}_z v^{-1}$, uniform for $|\xi| > \rho$. So, using equation (3.9.25) and its x derivative, we obtain

$$\tilde{v}_z \tilde{v}^{-1} = i\xi \begin{pmatrix} -1 & 0 \\ 0 & 1 \end{pmatrix} + i \left[C_1, \begin{pmatrix} -1 & 0 \\ 0 & 1 \end{pmatrix} \right]$$

$$= \begin{pmatrix} -i\xi & \phi \\ \phi^* & i\xi \end{pmatrix} .$$

(3.9.29)

A similar analysis can be performed near $\xi\zeta = 0$. By the asymptotic analysis of the eigenfunctions, it is possible to determine with the monodromy data and their various properties. Just as in the case of inverse scattering transform, it is possible to perform an inverse monodromy transformation to reconstruct the nonlinear field. The exhaustive analysis of the corresponding problem is outside the scope of the present discussions, and we refer to the works of Ablowitz and his collaborators. In the older literatures, there are exhaustive studies of Stoke's phenomena, monodromy matrices, and t-related problems. Although the above-mentioned method yields only an approximate form of the monodromy data, other approaches give reasonably good information about the data. One such method utilizes the theorem of Gurarii and Mateev.[46,68] The equation we consider is the three-wave interaction problem,

$$\frac{\partial u_1}{\partial t} + V_1 \frac{\partial u_1}{\partial x} = iqu_2^* u_3^*$$

$$\frac{\partial u_2}{\partial t} + V_2 \frac{\partial u_2}{\partial x} = iqu_1^* u_3^*$$

(3.9.30)

$$\frac{\partial u_3}{\partial t} + V_3 \frac{\partial u_3}{\partial x} = iqu_2^* u_1^* ,$$

and consider variables $\phi = \dfrac{x}{t}$; $\bar{u}_k = \dfrac{a_k}{u_k t} + b_k \dfrac{x}{t}$ where a_k, b_k are arbitrary constants. These new variables give rise to a single ordinary nonlinear equation of the form

$$(Z-1)^2 \Omega^2 = 1/z^2 \left(\phi^2 + \frac{1^2 - K^2 \phi^2}{Z(Z-1)^2} \right) \psi^2 ,$$

(3.9.31)

with

$$\Omega = \phi'' + \frac{3Z-1}{2Z(Z-1)} \phi' + \frac{2\phi I + \frac{1}{2} MI - \phi K^2}{Z(Z-1)^2}$$

$$I = \phi^2 + \frac{1}{2}\mu\phi + v$$

$$\psi = (z+1)\phi + \frac{1}{4}\mu(z+1) + \frac{\lambda}{2}(z-1) .$$

k,λ,μ,v are all constants defined as linear combinations of α,β_1, etc. Because this equation involves a square of the second derivative, it does not belong to the original Painleve class of equations. But it was demonstrated by Fokas and Yortos[67] that this equation can be connected to Painleve VI equation via a complicated set of transformations. But we can analyze the monodromy property of this equation using the technique due to Flaschka and Newell.[103] The linear problem associated with equation (3.9.30) is

$$i\psi_x = \left(\lambda A^{-1} - A^{-1}[A,Q]\right)\psi$$

$$\psi_t = iB\psi_x + [B,Q]\psi ,$$

(3.9.33)

where

$$A = \text{diag}(a_1 \cdot a_2, a_3)$$

and

$$Q = \begin{pmatrix} 0 & pa_{12}^{-1/2}u_2 & pa_{13}^{-1/2}u_1^* \\ pa_{12}^{-1/2}u_2^* & 0 & pa_{23}^{-1/2}u_3 \\ -pa_{13}^{-1/2}u_1 & pa_{12}^{-1/2}u_3^* & 0 \end{pmatrix}$$

(3.9.34)

$$B = \text{diag}(b_1,b_2,b_3) .$$

If we also set $\zeta = \lambda t$, the Lax equations are converted to the following linear equations in similarity variables:

$$\frac{id\psi}{dz} = \begin{pmatrix} \zeta/a_1 & \dfrac{pa_{21}^{1/2}a_2}{z_2 a_1} & -\dfrac{pa_{23}^{1/2}}{z_1^*}\dfrac{a_1^*}{a_1} \\ p\dfrac{a_{12}^{1/2}a_2^*}{z_2^* a_2} & \zeta/a_2 & -\dfrac{pa_{13}^{1/2}}{z_3}\dfrac{a_3}{a_2} \\ -p\dfrac{a_{13}^{1/2}a_1}{z_1 a_3} & \dfrac{pa_{23}^{1/2}}{z_3^*}\dfrac{a_3^*}{a_3} & \zeta/a_3 \end{pmatrix} .$$

(3.9.35)

Where we have used the notation

$$a_{ij} = a_i - a_j \; ;$$
$$z_i = u_i - b_i z , \qquad z_i^* = u_i^* - b_i^* z \qquad (i = 1, 2, 3) ,$$

we also have

$$\frac{d\psi}{d\zeta} = \left(\Sigma_{ij} \right) \psi \tag{3.9.36}$$

with

$$\Sigma_{11} = \frac{b_1 - iz}{a_1} , \qquad \Sigma_{22} = \frac{b_2 - iz}{a_2} , \qquad \Sigma_{33} = \frac{b_3 - iz}{a_3}$$

$$\Sigma_{12} = \frac{1}{\zeta} \left[p a_{12}^{-1/2} \left(\frac{b_1 a_2}{a_1} - b_2 \right) \frac{a_2}{z_2} + \frac{izp a_{12}^{1/2}}{z_2} \frac{a_2}{a_1} \right]$$

$$\Sigma_{21} = \frac{1}{\zeta} \left[p a_{12}^{-1/2} \left(\frac{a_1 b_2}{a_2} - b_1 \right) \frac{a_2^*}{z_2^*} - \frac{izp a_{12}^{1/2}}{z_2^*} \frac{a_2^*}{a_1} \right] ,$$

with similar expressions for the other matrix elements. To study the asymptotic behavior in the ζ-plane, we rewrite equation (3.9.36) as

$$\frac{d\psi}{d\zeta} = \begin{pmatrix} a & b/\zeta & c/\zeta \\ d/\zeta & e & f/\zeta \\ g/\zeta & h/\zeta & k \end{pmatrix} \psi = T\psi . \tag{3.9.37}$$

Let ψ_0 satisfy

$$\frac{d\psi_0}{d\zeta} = \begin{pmatrix} a & & \\ & e & \\ & & k \end{pmatrix} \psi_0 , \tag{3.9.38}$$

that is

$$\psi_0 = \begin{pmatrix} e^{a\zeta} & & \\ & e^{e\zeta} & \\ & & e^{k\zeta} \end{pmatrix} ,$$

and set $\psi = \psi^0 \phi$, then

$$\phi_\zeta = \left(\psi^0\right)^{-1} T\left(\psi^0\right)\phi$$

$$= \begin{pmatrix} 0 & b / \zeta e^{-(a-e)} & c / \zeta e^{-(a-k)} \\ d / \zeta e^{-(e-a)} & 0 & f / \zeta e^{-(e-k)} \\ g / \zeta e^{-(k-a)} & h / \zeta e^{-(k-c)} & 0 \end{pmatrix} \phi \qquad (3.9.39)$$

$$= B\phi \ .$$

The eigenvalues of matrix B are the roots of the $\det|B - \lambda I| = 0$, which are approximately $\lambda = \dfrac{\lambda_1}{\zeta}, \ \dfrac{\lambda_2}{\zeta}, \ \dfrac{\lambda_3}{\zeta}$, where λ_1, λ_2, λ_3 are roots of

$$\begin{vmatrix} -\lambda & b & c \\ d & -\lambda & f \\ g & h & -\lambda \end{vmatrix} = 0 \ .$$

Now the matrix diagonalizing B is given as

$$\Lambda = \begin{pmatrix} 1 & \dfrac{\lambda_1 h + gb}{\lambda_1 g + dh} & \dfrac{\lambda_1^2 - bd}{\lambda_1 g + dh} \\[3mm] \dfrac{\lambda_2^2 - fh}{\lambda_2^b + ch} & 1 & \dfrac{\lambda_2 c + bf}{\lambda_2 b + ch} \\[3mm] \dfrac{\lambda_3 d + gf}{\lambda_3 f + cd} & \dfrac{\lambda_3^2 - gc}{\lambda_3 f + cd} & 1 \end{pmatrix} .$$

With the help of Λ, we find that

$$\psi \text{ near } \zeta \rightarrow 0 \approx \Lambda \begin{pmatrix} \zeta^{\lambda_1} e^{a\zeta} & 0 & 0 \\ 0 & \zeta^{\lambda_2} e^{e\zeta} & 0 \\ 0 & 0 & \zeta^{\lambda_3} e^{k\zeta} \end{pmatrix}$$

$$
= \begin{pmatrix}
\zeta^{\lambda_1} e^{a\zeta} & \zeta^{\lambda_2} e^{a\zeta} \dfrac{\lambda_1 h + gb}{\lambda_1 g + dh} & \zeta^{\lambda_3} e^{k\zeta} \cdot \dfrac{\lambda_1^2 - bd}{\lambda_1 g + dh} \\[2.5ex]
\zeta^{\lambda_1} e^{a\zeta} \cdot \dfrac{\lambda_2^2 - fh}{\lambda_2 b + ch} & \zeta^{\lambda_2} e^{e\zeta} & \zeta^{\lambda_3} e^{k\zeta} \dfrac{\lambda_2 c + bf}{\lambda_2 b + ch} \\[2.5ex]
\zeta^{\lambda_1} e^{a\zeta} \cdot \dfrac{\lambda_3 d + gh}{\lambda_3 f + cd} & \zeta^{\lambda_2} e^{a\zeta} \dfrac{\lambda_3^2 - cg}{\lambda_3 f + cd} & \zeta^{\lambda_3} e^{k\zeta}
\end{pmatrix} .
$$

For determining the behavior near $\zeta \to \infty$, we put $\eta = 1/\zeta$ and consider the situation near $p \to 0$. Equation (3.9.39) becomes

$$
\frac{d\psi}{d\eta} = \left(A/\eta^2 + B/\eta \right) \psi ,
$$

with

$$
A = \begin{pmatrix} a & 0 & 0 \\ 0 & e & 0 \\ 0 & 0 & k \end{pmatrix} \quad B = \begin{pmatrix} 0 & b & c \\ d & 0 & f \\ g & h & 0 \end{pmatrix} .
$$

Proceeding just as before, we get

$$
\psi \approx \begin{pmatrix}
\zeta^{-\lambda_1} \exp(\Sigma_{11}\zeta) & \zeta^{-\lambda_2} \exp(\Sigma_{22}\zeta)\Lambda_{12} & \zeta^{-\lambda_3} \exp(\Sigma_{33}\zeta)\Lambda_{13} \\[1.5ex]
\zeta^{-\lambda_1} \exp(\Sigma_{11}\zeta)\Lambda_{21} & \zeta^{-\lambda_2} \exp(\Sigma_{22}\zeta) & \exp(\Sigma_{33}\zeta)\Lambda_{23} \\[1.5ex]
\exp(\Sigma_{11}\zeta)\Lambda_{31} & \exp(\Sigma_{22}\zeta)\Lambda_{32} & \zeta^{-\lambda_3} \exp(\Sigma_{33}\zeta)
\end{pmatrix} . \qquad (3.9.41)
$$

We now consider the region $|z| < b_1 < b_2 < b_3$ with $1/2(j-1)\pi \le \arg \le 1/2 j\pi$, $j = 1, 2, \ldots$, whence we obtain the behavior of the eigenfunctions depicted in Table 3.1.

	v_3	v_2	v_1
$0 \le \arg \le 1/2\pi$	inc	inc	dec
$\frac{1}{2}\pi \le \arg \le \pi$	dec	dec	inc
$\pi \le \arg \le 3/2\pi$	dec	dec	inc
$\frac{3}{2}\pi \le \arg \le 2\pi$	inc	inc	dec
$2\pi \le \arg \le 2\pi + \delta$	inc	inc	dec

Table 3.1

From this estimate of these asymptotics, we can assert that the solutions in different sectors are connected by matrices as

$$\psi^{(i)} = T\psi^{(i+1)} ,$$

where $\psi^{(i)}$ represents the solution in the i-th sector and $T = \left(T_{ij}\right)$ are Stoke's parameters. For the explicit determination of the Stoke's parameters, we can follow an elegant formulation of Gurarii and Mateev for the case of $n \times n$ coupled linear differential equations.

Theorem: Suppose there is a set of ordinary differential equations

$$t^{-q} \frac{du}{dt} = A(t)u \tag{3.9.42}$$

q is a non-negative integer, and $A(t)$ is a $n \times n$ complex matrix. Satisfying the following conditions,

(1) $A(t)$ is analytic in a neighborhood of infinity of the complex t plane and $A(t) = A_0 + A_1 t + \ldots$.

(2) The eigenvalues $\lambda_0, \lambda_1, \ldots, \lambda_n$ of A_0 are all different.

(3) The matrices $A_0, A_1, \ldots, A_m$ are all diagonalizable.

If we write

$$A(t) = B(t) + \tilde{A}(t)$$

$$B(t) = \overset{q+1}{\underset{p=1}{M}} A_p t^{-p}$$

$$\mu_k(t) = \lambda_k \frac{t^{q+1}}{q+1} + \lambda_k^{(1)} \frac{t^q}{q} + \ldots + \lambda_k^q t + \lambda_k^{q+1} \ln t \quad (k = 1, \ldots, n)$$

and set $V(t) = \mathrm{diag}\!\left(e^{\mu_1(t)}, e^{\mu_2(t)}, \ldots, e^{\mu_n(t)}\right)$, then the Stoke's multiplier matrix T_{ik}^1 is determined by the integral

$$T_{ik}^1 = \int_{\Gamma_{ik}^1} e^{-\mu(t)} \left(\tilde{A}(t) u^{(1)}(\tau) \right)_{ik} \tau^q d\tau \tag{3.9.43}$$

where $(\ )_{ik}$ denote (ik)th element.

To apply this theorem, we observe that in the case $q = 1$, the matrix on the right side is to be diagonalized not only to leading order but exactly. So a refined calculation leads to the following result:

$$\lambda_1 = \alpha_1/\eta + \beta_1/\eta^2, \quad \lambda_2 = \alpha_2/\eta + \beta_2/\eta^2, \quad \lambda_3 = \alpha_3/\eta^2 + \beta_3/\eta^2,$$

where $\alpha_1, \alpha_2, \alpha_3$ are roots of $x_0^3 + bgf + cdh - x_0(bd + cg + hf) = 0$, and $\beta_1, \beta_2, \beta_3$ are roots of

$$x_1^3 + x_1^2(a + e + k) + x_1(ae + ek + ak) + ake = 0.$$

The eigenvalues $\mu_k(t)$ of equation (3.9.43) turns out to be

$$\mu_1(\eta) = \tfrac{1}{2}\alpha_1\eta^2 + \beta_1\eta$$

$$\mu_2(\eta) = \tfrac{1}{2}\alpha_2\eta^2 + \beta_2\eta$$

$$\mu_3(\eta) = \tfrac{1}{2}\alpha_3\eta^2 + \beta_3\eta,$$

so the Stoke's parameter can be obtained from equation (3.9.43). In the present case,

$$T_{ik} = \sum_{v=0} S_v r^v \left(\sum_{j=1}^{v} A_j + q + 2C_{r-j} \right), \tag{3.9.44}$$

with

$$S_v = \int \exp(k(\tau) - \mu_1(\tau)) \tau^{-v-2} d\tau \quad v = 0, 1, 2, \dots.$$

To give an idea of the computations involved, we consider

$$S_{12} = \int e^{\mu_1 - \mu_2} \tau^{-2} d\tau$$

$$= \int \exp\left[\tfrac{1}{2}\eta^2(\alpha_1 - \alpha_2) + (\beta_1 - \beta_2)\eta\right] \eta^{v-2} d\eta.$$

But it is known that

$$\int_0^\infty x^{\nu-1}e^{-\beta x^2-x}dx=(2\beta)^{-1/2}\Gamma(\nu)\exp\left(\nu^3/8\beta\right)D_{-\nu}\left(r/\nu_2\beta\right),$$

so that

$$S_{12}=[\alpha(\omega-1)]^{-1/2}\Gamma(\nu-1)\exp\left(\frac{2j(ake)^3}{(a+k+e)^3(\omega-1)}\right)D_{-\nu+1}(\theta)$$

$$\theta=-\frac{2j\ ake}{(a+k+e)(\omega-1)}$$

and D_ν is the parabolic cylindrical functions.

Our above discussions point to the fact that the Painleve equations obtained as reductions of integrable partial differential equations can be analyzed with respect to its monodromy properties in the eigenvalue plane, and the explicit (sometimes approximate) form of the Stoke's parameter can be obtained. On the other hand, Ablowitz and others have done extensive work regarding the formulation of the Inverse monodromy problem, but the elaborate discussion of this problem is outside the scope of this discussion, and the interested reader is referred to the original papers for more information.

Chapter 4

Discrete and some special systems

4.1 Painleve test and discrete systems

An important class of nonlinear equations that occur in various branches of natural sciences are the discrete nonlinear systems. We have already discussed such systems in relation to ordinary differential equations (ODEs) that govern the behaviors of dynamical systems. Some new complications arise when we have n-coupled equations or when the time variable is also discrete, that is we have a mapping instead of a differential equation. This chapter sorts out the various intricacies associated with such nonlinear systems and their relation to the discrete Painleve equations. One of the most commonly known nonlinear equations is the Toda Lattice system. Let us start our discussion with this system and see how the different features of the Painleve test in the discrete system unfold in stages.

Toda Lattice equation

The equation can be written as

$$\ddot{Q}_n = \exp(Q_{n-1} - Q_n) - \exp(Q_n - Q_{n+1}) \ . \tag{4.1.1}$$

If we define variables $q_n = Q_n$ and $p_n = \exp(Q_n - Q_{n+1})$, then it can be rewritten as

$$\dot{p}_n = p_n(q_n - q_{n-1})$$
$$\dot{q}_n = p_{n-1} - p_n \ . \tag{4.1.2}$$

Originally, Toda obtained solution of the system in the form

$$Q_n = S_{n-1} - S_n \ ,$$

so the equation can be rewritten as

$$\varepsilon + \ddot{S}_n = \exp(S_{n+1} + S_{n-1} - 2S_n) \ ,$$

where $\varepsilon = 1$ for pure solitons. Toda chose to set

$$S_n = \log f_n \ ,$$

whence

$$\varepsilon + \frac{\partial^2}{\partial t^2} \log f_n = \frac{f_{n+1} f_{n-1}}{f_n^2} \ . \tag{4.1.3}$$

The variables p_n, q_n can also be connected to f_n by the equation

$$p_n = \frac{\partial^2}{\partial t^2} \log f_n \ .$$

The original Painleve analysis of the Toda system was done by Bountis, Vivaldi, and Segur for only three particles. They tried to obtain local Laurent expansion for each dependent variable about a common pole position to. For example, if we set

$$f_n(t) = g_n (t - t_0)^{-n^2} \ ,$$

then from equation (4.1.3) if $\varepsilon = 0$, g_n will satisfy

$$g_{n+1} g_{n-1} = n^2 g_n^2 \ .$$

If $g_0 = 1$ and $g_{-1} = 0$, then this can be solved by

$$g_n = (n-1)^2 (n-2)^4 (n-3)^6, ..., 2^{2(n-2)} g_1^n \ ,$$

which is consistent with the leading order analysis. But the approach of Bountis, Vivaldi, and Segur becomes increasingly complicated so Gibbon and Tabor introduced a set of Singular manifolds $\{\phi_n(t)\}$ like that of continuous case. It is possible then to make the connection with the Inverse Scattering Transform (IST) framework rather than identifying single-valued local Laurent expansions. In analogy with the truncated expansion of Weiss, Tabor, and Carnavel (WTC), we set

$$p_n = p_n^0/\phi_n^2 + p_n^1/\phi_n + p_n^{(2)}$$

$$q_n = q_{n-1}/\phi_{n-1} + q_n^0/\phi_n + q_n^{(1)}$$

(4.1.4)

and demand that both (p_n, q_n) and $\left(p_n^2, q_n^1\right)$ satisfy the same equation. Substituting equation (4.1.4) in equation (4.1.2) and equating various powers of ϕ_n^{-i}, we get

$$\phi_n^{-3} : q_n^0 = \dot{\phi}_n ; \qquad \phi_n^{-2} : p_n^0 = -\dot{\phi}_n^2 ,$$

(4.1.5)

and the remnant leads to $p_n^{(1)} = \ddot{\phi}_n$ use of equation (4.1.5), and the rest of equations yield

$$\frac{\partial^3}{\partial t^3} \log \phi_n = \left(\frac{\partial^2}{\partial t^2} \log \phi_n\right)\left(\frac{\dot{p}_n^{(2)}}{p_n^{(2)}} + \frac{\dot{R}}{R}\right) + p_n^{(2)} \frac{\dot{R}}{R} ,$$

(4.1.6)

where

$$R = \frac{\phi_{n+1}\phi_{n-1}}{\phi_n^2} .$$

Let us set

$$L = \frac{\partial^2}{\partial t^2} \log \phi_n ,$$

whence equation (4.1.6) can be put in the form

$$\frac{\partial}{\partial t}\left(L/\overline{p}_n\right) = \dot{R}/R\left(1 + L/p_n^{(2)}\right) ,$$

which integrates to

$$\frac{\partial^2}{\partial t^2} \log \phi_n = p_n^{(2)}\left[\frac{\phi_{n+1}\phi_{n-1}}{\phi_n^2} - 1\right] .$$

On the other hand, equations (4.1.4) and (4.1.5) become

$$p_n = \frac{\partial^2}{\partial t^2} \log \phi_n + p_n^{(2)}$$

$$q_n = \frac{\partial}{\partial t} \log \frac{\phi_{n-1}}{\phi_n} + q_n^{(1)}$$

which is the Backlund transformation. It may be remarked that the same procedure is also applicable even in the case of a two-dimensional Toda system.

$$Q_{n,xt} = \exp(Q_{n-1} - Q_n) - \exp(Q_n - Q_{n+1})$$

the corresponding form of coupled equations turns out to be,

$$p_{n,x} = p_n(q_n - q_{n+1})$$

$$q_{n,t} = p_{n-1} - p_n \ ,$$

with $q_n = Q_{n,x}$. The corresponding equation for ϕ_n becomes

$$\frac{\partial^2}{\partial x \partial t} \log \phi_n = p_n^{(2)} \left[\frac{\phi_{n+1} \phi_{n-1}}{\phi_n^2} - 1 \right]$$

or

$$\phi_n \phi_{n,xt} - \phi_{n,x} \phi_{n,t} = p_n^2 \left(\phi_{n+1} \phi_{n-1} - \phi_n^2 \right) \ . \tag{4.1.7}$$

Guibbon *et al.* proved that the functions ϕ_n are related to the eigenfunctions of Korteweg-de Vries (KdV) equation, so we seek two equations of the following form:

$$\phi_{n,x} = \alpha_{n+1} \phi_n + \beta_n \phi_{n+1}$$

$$\phi_{n,t} = \gamma_n \phi_{n-1} + \delta_{n-1} \phi_n \ . \tag{4.1.8a}$$

When these are used in equation (4.1.7), we get

$$\phi_n^2 (\alpha_{n+1,t} + \beta_n \gamma_{n+1}) + \phi_n \phi_{n+1} (\beta_n \delta_n - \beta_{n,t} - \beta_n \delta_{n-1})$$

$$- \phi_{n+1} \phi_{n-1} \beta_n \gamma_n = p_n^{(2)} \left(\phi_{n+1} \phi_{n-1} - \phi_n^2 \right) \ . \tag{4.1.8b}$$

Equating various powers of ϕ_n

$$\alpha_{n+1,t} + \beta_n \gamma_{n+1} = -p_n^{(2)}$$
$$\beta_n \gamma_n = -p_n^{(2)} \qquad\qquad (4.1.9)$$
$$\beta_{n,t} = \beta_n(\delta_n - \delta_{n-1}) \, .$$

Also, upon cross-differentiating, equations (4.1.8) and (4.1.9) yield

$$\frac{\gamma_{n,x}}{\gamma_n} = \alpha_{n+1} - \alpha_n$$
$$\alpha_{n+1,t} + \beta_n \gamma_{n+1} = \gamma_n \beta_{n-1} + \delta_{n-1,x} \, .$$

Our main contention is to choose $\alpha_n, \beta_n, \gamma_n$ and δ_n so that these are satisfied. Suppose we let $\beta_n = \text{constant} = \lambda$, so that

$$\gamma_n = -\lambda^{-1} p_n^{(2)} \quad \text{and} \quad \alpha_n = -q_n^{(1)}, \quad \delta_n = 0 \, .$$

The linear problem then becomes

$$\phi_{n,x} = -q_n^{(1)} \phi_n + \lambda \phi_{n+1}$$
$$\alpha_{n,t} = -\lambda^{-1} p_n^{(2)} \phi_{n-1} \, . \qquad\qquad (4.1.10)$$

Hence we can interpret λ as the spectral parameter and ϕ_n as the eigenfunction. It may be noted that equation (4.1.10) can also be derived from Hirota's approach.

Discrete Schwarzian derivative
Previous discussions repeatedly emphasized the importance of the Schwarzian derivative in the Painleve test. This explicitly displays the invariance of the whole procedure under conformal transformation. Let us recapituate the basic form of the Schwarzian derivative:

$$\{\phi, x\} = \phi_{xxx}/\phi_x - 3/2(\phi_{xx}/\phi_x)^2$$
$$= (\phi_{xx}/\phi_x)_x - 1/2(\phi_{xx}/\phi_x)^2 \, .$$

Note that if we set $v = \phi_{xx}/\phi_x$, it can be rewritten as

$$\{\phi, x\} = v_x - 1/2v^2 \ ,$$

a kind of Miura map, transforming KdV to an MKdV equation. The important question that now arises is, is it possible to obtain a similar quantity in case of discrete equations. We again refer back to equation

$$\phi_n \phi_{n,xt} - \phi_{nt}^2 = p_n^{(2)}\left(\phi_{n+1}\phi_{n-1} - \phi_n^2\right) ,$$

which is nothing but (4.1.7). We can split this into two equations:

$$\phi_{n,tt} = p_n^{(2)}\left(\phi_{n+1} + \phi_{n-1} - 2\phi_n\right) \tag{4.1.11}$$

$$\phi_{n,t}^2 = p_n^{(2)}\left\{\phi_n\left(\phi_{n+1} + \phi_{n-1}\right) - \phi_n^2 - \phi_{n+1}\phi_{n-1}\right\} .$$

Dividing these, we get

$$\frac{\phi_{n,tt}}{\phi_{n,t}^2} = \frac{\phi_{n+1} + \phi_{n-1} - 2\phi_n}{\phi_n\left(\phi_{n+1} + \phi_{n-1}\right) - \phi_n^2 - \phi_{n+1}\phi_{n-1}} .$$

By a laborious computation, one can verify that the expression on the right side is invariant under a general fractional transformation. Therefore, one can call this the discrete Schwarzian.

Before passing to the generalization of the Toda Lattice, first look at the three-particle free-end Toda chain following the original approach of Bountis, Segur, and Vivaldi. The Hamiltonian is

$$H = p_1^2/2m_1 + p_2^2/2m_2 + p_3^2/2 + e^{\varepsilon(q_1 - q_2)} + e^{q_2 - q_3} , \tag{4.1.12}$$

where we have normalized m_3 to 1. We now make a change of variable

$$\dot{a}_1 = \frac{1}{2}e^{\varepsilon(q_1 - q_2)/2}$$

$$\dot{a}_2 = \frac{1}{2}e^{(q_2 - q_3)/2} , \qquad b_k = p_k/2mk .$$

The equations of motion are

$$\dot{a}_1 = \varepsilon a_1(b_2 - b_1), \quad \dot{a}_2 = -a_2\big[(1+m_2)b_2 + m_1 b_1\big]$$

$$\dot{b}_1 = 2\varepsilon/m_1\, a_1^2, \quad \dot{b}_2 = 2/m_2\big(a_2^2 - \varepsilon a_1^2\big), \quad m_1, \quad m_2, \quad \varepsilon > 0. \tag{4.1.13}$$

We search for a_1, a_2 in the form $a_1 \sim c_1(t-t_0)^p$, $a_2 \sim c_2(t-t_0)^q$ near $t - t_0$. One observes that there are three choices:

(i) $a_1 \sim c_1\tau^p + ...$, $\quad a_2 \sim c_2\tau^{-1}$, $\quad p > -1$

(ii) $a_1 \sim c_1\tau^{-1}$, $\quad a_2 \sim c_2\tau^q$, $\quad q > -1$

(iii) $a_1 \sim c_1\tau^{-1}$, $\quad a_2 \sim c_2\tau^{-1}$,

as $\tau \to 0$ with $\tau = t - t_0$. Substituting these expressions in equation (4.1.13) and comparing lead order coefficients, we get

$$a_1 \sim c_1\tau^p, \quad a_2 \sim i\big[m_2/2(1+m_2)\big]^{1/2}\tau^{-1} + ...$$

$$b_1 \sim 2c_1^2\varepsilon/(2p+1)m_1 \cdot \tau^{2p+1} \tag{4.1.14}$$

$$b_2 \sim 1/(1+m_2)\tau.$$

Therefore, dominant behavior does not introduce any branch point. Here c_1 is an arbitrary constant and

$$p = \frac{\varepsilon}{1+m_2} 2p = \text{integer} > 0.$$

The resonance equation turns out to be

$$\gamma(2p+1+\gamma)(\gamma+1)(\gamma-2) = 0, \tag{4.1.15a}$$

so the system will have Painleve property if there are three arbitrary constants, t_0, c_1 and the constant entering at $\gamma = 2$. In the second case, the resonance equation turns out to be $(\gamma+1)\gamma(\gamma-1)(\gamma-2) = 0$, and we have four arbitrary constants t_0, c_2, f and g entering at $\gamma = -1, 0, 1$, and 2, respectively. The expansions are

$$a_1 \cong \frac{i}{e}\left(m_1 m_2/2(m_1+m_2)^{1/2}\right)\left(\tau^{-1}+g\tau+...\right)$$

$$a_2 \sim c_2\tau^q\left(1-\varepsilon f/2-\varepsilon\cdot\tau+...\right)$$

$$b_1 \sim \frac{m_2}{M\varepsilon(m_1+m_2)}\tau^{-1}+f-\frac{2m_2 g}{\varepsilon(m_1+m_2)}\tau+... \tag{4.1.15b}$$

$$b_2 \sim \frac{-m_1}{\varepsilon(m_1+m_2)}\tau^{-1}+f+\frac{2m_1 g}{\varepsilon(m_1+m_2)}\tau+... \ .$$

In both cases we can ascertain that the three-particle Toda Lattice with free-end has only movable pole type singularity. The treatment is clearly not suitable for extension to an n-particle system. As the third situation is concerned, the resonance equation is

$$\gamma^2 - \gamma - 2M = 0 \tag{4.1.16}$$

with

$$M = \frac{(1+p)(1+q)}{1-pq}$$

$$p = m_1 + \varepsilon(m_1+m_2), \quad q = 1+m_2+\varepsilon \ .$$

For the Painleve property, one must have integer roots only, and this implies

$$M = \frac{(1+p)(1+q)}{1-pq} = \frac{n(n+1)}{2} \quad n = 0, 1, 2,..., $$

with $+2p$ and $2q$ positive integers.

(a) There exists many possible solutions such as $p = q = 1/2$, $n = 2$, and resonances at $\gamma = -2, -1, 2, 3$. In this case,

$$m_1 = \frac{\varepsilon(2-1)}{2-\varepsilon} \quad m_2 = 2\varepsilon - 1 \ .$$

(b) There is the situation

$$p = 1/2, \quad q = 1, \quad \text{or} \quad p = 1, \quad q = 1/2 \quad \text{for} \quad n = 3$$

and resonances at $\gamma = -3, -1, 2, 4$. In this case,

$$m_1 = \frac{\varepsilon(\varepsilon - 1)}{2 - \varepsilon} \qquad m_2 = \varepsilon - 1 \ .$$

Our above discussion elaborates the various approaches to the discrete Painleve analysis in the case of Toda Lattice. Toda Lattice has been generalized by associating the particles with the root vectors of Lie algebra. But there exists various kinds of Lie algebra, and recently it has been found that if one constructs a Toda Lattice by involving a Hyperbolic Lie algebra, then that system is no longer completely integrable. Such a generalization of the Toda system was done by Gebert *et al.*, and it is different from equation (4.1.1). This new equation is written as

$$\partial_+ \partial_- \phi_i = \exp\left(\beta \sum_{j=1}^{N} K_{ij} \phi_j \right), \tag{4.1.17}$$

where K_{ij} is the generalized Cartan matrix of a Kac-Moody algebra. It is in general a $N \times N$ matrix. It satisfies few basic conditions,

(a) $K_{ij} \varepsilon Z$

(b) $K_{ii} = 2$

(c) $K_{ij} \leq 0 (i \neq j)$

(d) $K_{ij} = 0$ if $K_{ij} = 0$.

In particular, the Lie algebra may be a hyperbolic Kac-Moody algebra. Many authors have studied the properties of such generalized Toda Lattice. Recently, Gebert *et al.* analyzed the Painleve property of such system. They have showed that these systems have resonance only at $n = 2$ and therefore are not completely integrable. We can rewrite equation (4.1.17) as

$$\frac{\partial A_i}{\partial t} = A_i \sum_{j=1}^{N} K_{ij} B_j \quad (i = 1, ..., N)$$

$$\frac{\partial B_j}{\partial x} = A_j = -\sum_{i=1}^{N} \delta_{ij} A_i \quad (j = 1, ..., N) \ .$$

Suppose that $\phi(x, t) = 0$ is the singular manifold. Then we set

$$A_i = \phi^{-n_A} \sum_{n=0}^{\infty} \phi^n A_i^{(n)}, \qquad B_i = \phi^{-n_B} \sum_{n=0}^{\infty} B_j^{(n)} \phi^n$$

in the above coupled set and obtain

$$T^{(n)} \bar{X}^{(n)} = \bar{b}^{(n)} \tag{4.1.18}$$

for the expansion coefficients

$$\bar{X}^{(n)} = \left(A_1^{(n)}, ..., A_N^{(n)}, B_1^{(n)}, ..., B_N^{(n)} \right)^T,$$

where $T^{(n)}$, $\bar{b}^{(n)}$ are given by

$$T^{(n)} = \begin{pmatrix} P^{(n)} & Q^{(n)} \\ R^{(n)} & S^{(n)} \end{pmatrix} \qquad \bar{b}^{(n)} = \begin{pmatrix} b_i^{(n)} \\ b_{n+j}^{(n)} \end{pmatrix}$$

$$P_{ik}^{(n)} = \left\{ (n-2)\dot{\phi} - \sum_{j=1}^{N} K_{ij} B_j^{(0)} \right\} \delta_{ik}$$

$$Q_{il}^{(n)} = -A_i^{(0)} K_{il}$$

$$R_{jk}^{(n)} = \delta_{jk}$$

$$S_{jl}^{(n)} = (n-1)\phi_x \delta_{jl}$$

$$b_i^{(n)} = -\dot{A}_i^{(n-1)} + \sum_{m=1}^{n-1} A_i^{(n-m)} \cdot \sum_{j=1}^{N} K_{ij} B_j^{(m)}$$

$$b_{N+j}^{(n)} = -B_{jx}^{(n-1)},$$

where
$$i,j,k,l = 1,...,N, \quad \text{and} \quad n_A = n_{B+1}.$$

For strictly hyperbolic Koc-Moody algebras,

$$n_A = 2, \quad n_B = 1.$$

For the resonance position, $\det T^{(n)}$ must be 0 and from (4.1.18)

$$B_i^{(0)} = -2\dot\phi \sum_{j=1}^{N} \left(K^{-1}\right)_{ij}, \quad A_i^{(0)} = -2\dot\phi\phi_x \sum_{j=1}^{N} \left(K^{-1}\right)_{ij} \quad l = 1,...,N .$$

The matrix $T^{(n)}$ can then be written as

$$T^{(n)} = \begin{pmatrix} n\dot\phi & -A_1^0 K_{11} & & -A_1^0 K_{1N} \\ & 0 & & \\ 0 & & & \\ & n\dot\phi - A_N^0 K_{N1} & & -A_N^0 K_{NN} \\ 0 & (n-1)\phi_x & 0 & \\ 0 & 1 & 0 & (n-1)\phi_x \end{pmatrix}.$$

It is then easy to see that $\det T^{(n)}$ is given by

$$\det T^{(n)} = \left(\dot\phi\phi_x\right)^N \det[n(n-1)\cdot 1 - 2DK] , \tag{4.1.19}$$

where

$$D = \begin{bmatrix} \sum_{j=1}^{N} \left(K^{-1}\right)_{ij} & \cdots & \\ & \sum_{j=1}^{N} \left(K^{-1}\right)_{Nj} & \end{bmatrix}.$$

Gebert *et al.* evaluated this for different Lie algebra and obtained the following forms for the other factor of $\det T^{(n)}$.

Lie algebra

$$AN \qquad : (\lambda - 1, 2)(\lambda - 2, 3)(\lambda - 3\cdot 4)...(\lambda - N(N+1))$$
$$BN(CN) : (\lambda - 1\cdot 2)(\lambda - 3\cdot 4)...(\lambda - (2N-1)2N)$$
$$G_2 \qquad : (\lambda - 2)(\lambda - 30) \tag{4.1.20}$$
$$F_4 \qquad : (\lambda - 2)(\lambda - 30)(\lambda - 56)(\lambda - 132), \text{ etc.}$$

where
$$\lambda = n(n+1) .$$

The number appearing in each factor is always a product of two consecutive integers (e.g., $20 = 4 \times 5$, $56 = 7 \times 8$ and 50 on). The larger one is a possible resonance.

Remarkably, it is found that resonances occur precisely at the values

$$n = \text{exponents} + 1 .$$

The compatibility can also be checked for simple Lie algebras. The important point to note is that the resonances occur not $2N$ times, but N times where N is the rank. Therefore, we have fewer resonances. On the other hand, it is known that the values of conserved W currents (currents generating the W algebra) and the exponents of Lie algebra correspond one to one, so it is important to note that the Painleve test not only gives us information about the integrability, but also yields detailed information about the existing conserved current.

Integrable mappings

Discrete nonlinear systems such as the Toda Lattice are common examples of ordinary dynamical systems. In recent years, the relevance of nonlinear mappings have been felt after the famous discovery of Fiegenbam in relation to logistic map. This type of nonlinear mapping may arise when both the space and time variables in a nonlinear PDE are discretized. For example, the mapping analogue of potential KdV is

$$x_j^{i+1} = x_{j+1}^{i-1} + 1 \big/ x_j^i - 1 \big/ x_{j+1}^i , \tag{4.1.21}$$

where i and j are two indices varying over a two-dimensional lattice given below

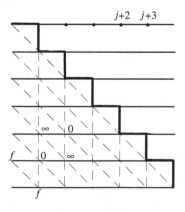

The index j runs over the vertical lines; i refers to slanted lines. Initial data is given on the heavily drawn segments. The symbols 0, ∞, and f indicate the location of the corresponding values on the lattice. Evolution can be understood as taking place toward

increasing i.   This was the idea put forward by Grammaticos, Ramani, and Papageorgiou.[79]   Now it may happen that by starting with some initial value of x, the iteration by equation leads to a value 0 at some (i, j). This is quite possible, and the point where it occurs depends on the initial data that is the singularity induced is movable. Now equation (4.1.21) asserts that x diverges both at $(i+1, j-1)$ and $(i+1, j)$ and vanishes at $(i+2, j-1)$. Now an important question is what happens at the sites $(i+3, j-2)$ and $(i+3, j-1)$. It is observed that a very fine cancellation of finite values takes place at these locations, thus the singularity is perfectly confined. The whole phenomena form a dynamics of the map. So, one may assert that integrability in mapping systems means that singularities are confined. The concept of Singularity Confinement can be illustrated with the help of a discretized anharmonic oscillator, which was initially presented by Hirota,

$$\left[x_{n+1} - 2x_n + x_{n-1}\right]/\delta^2 = -ax_n - \frac{\beta}{2} x_n^2\left[x_{n+1} + x_{n-1}\right] . \tag{4.1.22}$$

Yoshida proved that this mapping is the symplectic integrator of order 2 for the quartic oscillator. The complete solution can be written as

$$x(t) = x_0 cn[\Omega(t - to), K]$$
$$1 - cn(\delta\Omega)/dn^2(\delta\Omega) = a\delta^2/2$$
$$2K^2 = x_0^2\beta\delta^2/[Sn(\delta\Omega)/dn(\delta\Omega)]^2$$
$$t = n\delta + to .$$

Assume that $nx(n)$ diverges. The addition formulae for the elliptic functions show that $x(n \pm 1) = \pm\left(2/\beta\delta^2\right)^{1/2}$ and $x(n+2) = -x(n-2)$. Thus, $x(n-1)$ has precisely the value that guarantees the divergence of $x(n)$, and $x(n+1)$ has the value that compensates the divergence. The whole idea about the integrability of the mapping tantamounts to the following fact: the integrable singularities of integrable mappings are confined, that is they are cancelled out after a finite number of steps.   One may also note that Hirota's anharmonic oscillator may be written as

$$Z_{n+1} + Z_{n-1} = 2\mu Zn/\left(1 + Z_n^2\right) , \tag{4.1.23}$$

where

$$Zn = \left(\frac{\beta\delta^2}{2}\right)^{1/2} x_n , \quad \mu = 1 - \alpha\delta^2/2 ,$$

which is the McMillan map. This map is a special case of

$$X_{n+1} + X_{n-1} = -\frac{ax_n^2 + bx_n + C}{dxn^2 + exn + f} . \tag{4.1.24}$$

One can apply the same interion as above to deduce constraints on the parameter $a_1 b_1, ..., f$ so that the mapping is integrable. There are two different but important situations: $d = 0$ and $f = 0$. The McMillan map is a special case of a more general situation discussed by Quispel, Roberts, and Thompson,[95]

$$x_{n+1} = \frac{f_1(x_n) - x_{n-1}f_2(x_n)}{f_2(x_n) - x_{n-1}f_3(x_n)} , \tag{4.1.25}$$

where f_i is a quartic polynomial. Integrability of mapping from the idea of monodromy was discussed by Papageorgiou, Nijhoff, and Capel.[94]

Compatibility and Painleve

The discrete Painleve equations have been studied only very recently. As in the case of a continuous system, it has been seen that integrable discrete systems lead to another type of nonlinear system known as discrete Painleve system. It was observed earlier that the usual Painleve equations can be obtained as the consistency condition of two equations

$$\begin{aligned} \frac{\partial v}{\partial x} &= Lv \\ \frac{\partial v}{\partial z} &= Mv \end{aligned} \tag{4.1.26}$$

where x is the space variable and z is the spectral parameter. The idea for the discrete system was adopted by Joshi *et al.* by starting with the equations

$$\frac{\partial v_n}{\partial z} = M_n(z)v_n \tag{4.1.27a}$$

$$v_{n+1} = L_n(z)v_n \ . \tag{4.1.27b}$$

v, L_n, v_n all depend on the continuous variable z and a discrete variable n. Let ψ_n be a fundamental matrix solution of equation (4.1.27) and N_n be the monodromy matrix around a singular point zo of the equation. N_n may be a Stoke's multiplier matrix if the singular point is irregular. If ϕ_n is the analytic continuation of ψ_n on a closed loop around z_0, then

$$\phi_n = \psi_n N_n \ . \tag{4.1.28}$$

An important proposition is that the invariance of the Monodromy matrix N_n under iteration of n is equivalent to the compatibility of equations (4.1.27) and (4.1.28).

Let z_0 be a regular singular point. If equations (4.1.27a) and (4.1.27b) are compatible, then

$$\frac{\partial L_n}{\partial z} + L_n M_n = M_{n+1} L_n \ .$$

Also, from equation (4.1.28)

$$\phi_{n+1} = \psi_{n+1} N_{n+1} \quad \text{and} \quad L_n \phi_n = L_n \psi_n N_{n+1} \ .$$

Using equation (4.1.28), again we get $N_n = N_{n+1}$. The compatibility condition is required to show that N_n are true monodromy matrices and they are independent of z. Differentiating equation (4.1.28) with respect to z gives us

$$M_n \phi_n = M_n \psi_n N_n + \psi_n \frac{\partial N_n}{\partial z} \ ,$$

which leads to

$$\frac{\partial N_n}{\partial z} = 0 \ .$$

Now start from the assumption $N_n = N_{n+1}$. Since $N_n = \psi_n^{-1} \phi_n$, we have

$$\phi_{n+1} = \psi_{n+1} N_n = \psi_{n+1} \psi_n^{-1} \phi_n$$

hence

$$\phi_{n+1} \phi_n^{-1} = \psi_{n+1} \psi_n^{-1} \ .$$

Let

$$L_n = \psi_{n+1}\psi_n^{-1}$$

so that

$$\frac{\partial L_n}{\partial z} = \frac{\partial \psi_{n+1}}{\partial z}\psi_n^{-1} - \psi_{n+1}\psi_n^{-1}\frac{\partial \psi_n}{\partial z}\psi_n^{-1}$$

$$= M_{n+1}L_n - L_n M_n \ . \tag{4.1.29}$$

Ablowitz and Ladik in their paper considered an L_n operator of the following form

$$L_n = 1/U_n \begin{pmatrix} z + R_n S_n & Q_n + S_n/z \\ ZT_n + R_n & Q_n T_n + 1/z \end{pmatrix} \tag{4.1.30}$$

$$U_n = 1 - S_n T_n \ ,$$

where the time part is

$$M_n = \begin{pmatrix} A_n(z) & B_n(z) \\ C_n(z) & D_n(z) \end{pmatrix} \tag{4.1.31}$$

and are rational in z. If we impose the compatibility between equations (4.1.30) and (4.1.31) with this particular form of M_n and L_n, we get (where $Q = 0$, $R = 0$, $T = 1$ and $S = 1 - \alpha$, $\lambda = z + z^{-1}$)

$$\frac{\partial \omega_n}{\partial \lambda} = a_n \omega_{n+1} + b_n \omega_n \ , \tag{4.1.32}$$

ω_n standing for second component of v_n. a_n and b_n are given by

$$a_n = \alpha_n C_n \frac{z}{z^2 - 1} \ , \qquad b_n = \frac{z^2 D_n - C_n}{z^2 - 1} \ ,$$

whence the compatibility conditions become

$$b_{n+1} - b_{n-1} + \lambda \left(\frac{a_{n+1}}{\alpha_{n+1}} - \frac{a_n}{\alpha_n} \right) = 0$$

$$\frac{\lambda^2}{\alpha_n}\left(\frac{a_{n+1}}{\alpha_{n+1}} - \frac{a_n}{\alpha_n}\right) + \frac{\lambda}{\alpha_n}(b_{n+1} - b_n) - \frac{1}{\alpha}\left(\alpha\frac{a_n}{\alpha_n} - \alpha_{n+1}\right) = 0 \ . \qquad (4.1.33)$$

Consider a_n and b_n to be polynomial in λ

$$a_n = A_0 + A_1\lambda^2$$
$$b_n = B_0 + B_1\lambda + B_2\lambda^2 + B_3\lambda^3 \ .$$

Therefore, equation (4.1.33) yields

$$A_0 = C_5\alpha_n + C_2\alpha_n(\alpha_n + \alpha_{n+1})$$
$$A_2 = C_2\alpha_n$$
$$B_0 = C_6$$
$$B_1 = -C_2\alpha_n + C_4$$
$$B_2 = C_3$$
$$B_3 = C_1$$

along with the condition

$$-C_7 + n = -C_5\alpha_n - C_2\alpha_n(\alpha_{n+1} + \alpha_n + \alpha_{n-1}) \ .$$

C_i are arbitrary constants.

Consider a combination of scaling and continuum limit

$$\alpha_n = 1 + \gamma u(nh) \ , \qquad \delta\xi = nh$$

and the equations obtained from consistency

$$(z + S_n R_n)(A_{n+1} - A_n) + (zT_n + R_n)_{n+1} - (Q_n + S_n/z)C_n = 1 \qquad (4.1.34)$$

$$(Q_n T_n + 1/z)B_{n+1} - (z + S_n R_n)B_n + (Q_n + S_n/z)(A_{n+1} - D_n) - S_n/z^2 \qquad (4.1.35)$$

$$(Z + S_n R_n)C_{n+1} - (Q_n T_n + 1/z)C_n + (zT_n + R_n)(D_{n+1} - A_n) = T_n \qquad (4.1.36)$$

$$(Q_n T_n + 1/z)(D_{n+1} - D_n) + (Q_n + S_n/z)C_{n+1} - (zT_n + R_n)B_n = -1/z^2 \ . \qquad (4.1.37)$$

From equations (4.1.34) and (4.1.37), we get

$$(Z + S_n R_n)(Q_n T_n + 1/z)(A_{n+1} - D_n + D_{n+1} - A_n) + (zT_n + R_n)$$
$$\times \left[(Q_n T_n + 1/z)B_{n+1} - (z + S_n R_n)B_n \right] + [Q_n + S_n/z]\{(z + S_n R_n)C_{n+1}$$
$$- (Q_n T_n + 1/z)C_n] = Q_n T_n - S_n R_n / z^2 \} .$$

And from (4.1.35) and (4.1.36), we get

$$(zT_n + R_n)(Q_n + S_n/z)(A_{n+1} - D_n + D_{n+1} - A_n) + (zT_n + R_n)$$
$$\times \left[(Q_n T_n + 1/z)B_{n+1} - (z + S_n R_n)B_n \right] + \left[(z + S_n R_n)C_{n+1} \right.$$
$$- (Q_n T_n + 1/z)C_n](Q_n + S_n/z) = Q_n T_n - S_n R_n / z^2 .$$

Subtracting these two equations gives us

$$(Q_n R_n - 1)(S_n T_n - 1)\left[(A_{n+1} - A_n) + (D_{n+1} - D_n) \right] = 0 .$$

So if $Q_n R_n \neq 1$, $S_n T_n = 1$, then

$$A_n = -D_n = d(z)$$

$d(z)$ independent of n.

An important case is given by the special situation $Q = 0$, $R = 0$, $T = 1$. If we write $S = 1 - \alpha$, $\lambda = z + 1/z$, equation (4.1.33) becomes

$$\alpha\omega- = \lambda\omega - \omega$$

being the second component of v. The second component of equation (4.1.33) yields

$$K_1 = \frac{\delta^3}{C_2 \gamma h^3} , \quad K_2 = \frac{\delta^2 \mu_1}{h^2} , \quad K_3 = -\frac{\delta^2 \mu_2}{h^2} , \quad K_4 = -3\frac{\gamma\delta^2}{h^2} , \qquad (4.1.38a)$$

where $h \to 0$ with $\delta = 0(h)$ gives

$$u_{\xi\xi} = K_4 u^2 + K_3 u + K_2 + K_1 \xi \ , \tag{4.1.38b}$$

which is nothing but P_1, the first Painleve. Therefore, we say equation (4.1.33) is the corresponding discrete Painleve equation.

On the other hand, if we set $d = 0$, that is $A = -D$, and substitute

$$A = K/z^3 + \left(\frac{2\mu f_n f_{n-1} + n + v}{z} \right) + az$$

$$B = -2K f_n / z^2 + 2\mu f_{n+1} \tag{4.1.39}$$

$$C = -2K f_n / z^2 + 2\mu f_{n-1} \ .$$

K, μ, v are constants. Then the equation governing f is

$$\mu f_{n+1} + K f_{n-1} + f_n \left[v + n + 1/2 - \mu f_n (f_{n+1} + f_{n-1}) \right] = 0 \ . \tag{4.1.40}$$

If we take $K = \mu \neq 0$, we can write this as

$$f_{n+1} + f_{n-1} = (K_1 n + K_2) f_n (1 - f_n^2) \ , \tag{4.1.41}$$

where

$$K_1 = -1/\mu \ , \qquad K_2 = K_1 (v + 1/2) \ .$$

Under the substitution

$$f_n = \gamma U(nh) \ , \qquad t = nh$$

and the scalings

$$K_1 = \mu_1 h^3 \ , \qquad K_2 = 2 + \mu_2 h^2 \ , \qquad \gamma = \mu_3 h \ ,$$

equation (4.1.41) gives

$$u_{tt} = (\mu_1 t + \mu_2) u + 2\mu_3^2 u^3 \ , \tag{4.1.42}$$

in the limit $h \to 0$

which is a different version of P_{11}. Therefore, equation (4.1.40) can be considered to be its discrete version.

A different approach was adopted by Ramani *et al.* They considered possible mapping equations considered to be discretized versions of the usual Painleve equation. As a concrete example, consider P_{1v}, which we know to be of the following form:

$$\omega^{11} = \omega^{12}/2 + 3\omega^3/2 + 4z\omega^2 + 2(z^2 - a)\omega + b/\omega \ .$$

In references (64), (78) the following mapping was considered

$$x_{n+1}x_{n-1} + x_n(x_{n+1} + x_{n-1})$$

$$= \frac{K(n)x_n^3 + \varepsilon(n)x_n^2 + \xi(n)x_n + \mu(n)}{x_n^2 + \beta(n)x_n + \gamma(n)} \tag{4.1.43}$$

as a discretized version of P_{1y}. Solving for x_{n+1}, we obtain two possible sources of divergence. Either z_n is a 0 of the denominator of $x_n + x_{n-1}$ vanishes. In the first case, starting with x_n, we find that x_{n+1} diverges and x_{n+2} has value $K(n+1) - x_n$. Employing $x_{n+3} = -x_{n+2}$ and to a divergent x_{n+4}, unless x_{n+2} is a 0 of $x^2 + \beta(n+2)x + \gamma(n+2)$. Expressing x_{n+2} in terms of x_n and demanding that the resulting equation is identical to $x_n^2 + \beta(n)x_n + \gamma(n)$, one obtains

$$\beta(n+2) + \beta(n) = -2K(n+1)$$

$$\gamma(n+2) + \beta(n+2)K(n+1) + K^2(n+1) = \gamma(n) \ ,$$

leading to $4\gamma(n) - \beta^2(n) = 4\gamma_0 \pm$. A similar analysis can be done for the second case.

Recently, Conte and Musette[60] considered a very pertinent question: what is the proper discretization procedure that preserves some global property and is explicit linearizability? A discrete equation can be considered a functional equation linking the values taken up by the dependent variable u at, say, $N+1$ points $x + kh$, $K2K_0 = 0$, N or xq^k, $k - K_0 = 0,..., N$. The integer N is called the order of the equation. They then considered a specific discretization procedure that was observed to preserve the Painleve property of the discrete system. An elegant prescription for the discretization was given recently by Ward. It is based on the idea of preserving the Leibniz rule for derivatives by using a nonlocal multiplication rule. Let x be an integer-valued single-discrete variable. Ward showed a multiplication rule of the form

$$(f \ g)(x) = \sum_{y,z} C_{xyz} f(y) g(z)$$

and a difference operator

$$\Delta f(x) = \sum_{y} dxy f(y)$$

$C_{xyz} d_{xy}$ are real constants, such that

(a) $*$ is commutative and associative,

(b) if f is constant, then $f*g = fg$

(c) if f is constant, $\Delta f = 0$

(d) the Leibniz rule

$$\Delta(f*g) = f*(\Delta g) + (\Delta f)* g$$

holds.

If f is periodic $f(x+N) = f(x)$, a solution is

$$(f*g)(x) = \langle f \rangle g(x) + \langle g \rangle f(x) - \langle f \rangle \langle g \rangle \langle f \rangle$$

being the average over N

$$\langle f \rangle = N^{-1} \sum_{x=1}^{N} f(x)$$

and

$$\Delta f(x) = f(x+1) - f(x).$$

In the Lax equation

$$\psi_{\xi} = A\psi; \quad \psi_n = B\psi$$

we then interpret $A\psi$ as $A*\psi$ and so on. It is then appropriate to consider variable 'ξ' as discrete and replace $\partial\xi$ by $\Delta\xi$ but treat 'η' as continuous. If we impose the consistency, we can get integrable discrete systems. The other difficulties of the present scheme are discussed in detail in references (78), (79), (80). The time discretization procedure was suggested by Kupershmidt and showed that these are also multi-Hamiltonian systems. Conte *et al.* observed that the two familiar Painleve equations (*PI*) and (*PII*),

$$PI : \ -d^2u/dx^2 + 6u^2 + x = 0$$

$$PII : \ -d^2u/dx^2 + 2u^3 + xu + \alpha = 0 \, ,$$

can be obtained as the $h \to 0$ of the following

$$-\frac{1}{h^2}\{v(x+h) - 2u(x) + v(x-h)\} + 2\{u(x+h) + u(x)$$

$$+ u(x-h)\}u(x) + x = 0 \qquad \text{(i)}$$

$$-\frac{1}{h^2}\{u(x+h) - 2u(x) + u(x-h)\} + (u(x+h) + u(x-h))$$

$$\times u^2(z) + xu(x) + \alpha = 0 \qquad \text{(ii)} \, . \qquad\qquad (4.1.44)$$

And for many reasons, these equations (i, ii) can be called discrete Painleve equations. Indeed, both admit discrete Lax pair,

$$\psi(x+h/2) = A\psi(x-h/2)$$

$$\partial z\psi(x-h/2) = B\psi(x-h/2) \, .$$

A and B are 2×2 matrices of the following form

$$A = \begin{pmatrix} 1/z & hu \\ hu & z \end{pmatrix}$$

$$t = z - z^{-1}/2h \; ; \quad z = 1 + ht + 0\left(h^2\right)$$

$$hzB = \left[1 - 2uu(x-h) + 4t^2 - (x-h/2)\right]\begin{pmatrix} 1 & 0 \\ 0 & -1 \end{pmatrix} + (z+1/z)$$

$$\times (u-u)(x-h)/h\begin{pmatrix} 0 & 1 \\ -1 & 0 \end{pmatrix} - [2t(u+u(x-h)) + \alpha/t]$$

$$\times \begin{pmatrix} 0 & 1 \\ 1 & 0 \end{pmatrix} \, ,$$

while the second is

$$A = \begin{pmatrix} 1-h^2z & h(2u-2t) \\ h & 1-h^2z \end{pmatrix}$$

$$t = z - h^2 z^2 / 2$$

$$z = t + \frac{t^2}{2}h^2 + 0\left(h^4\right)$$

$$B = \begin{pmatrix} 2\left(1-h^2z\right)(v-u(x-h))/h & \begin{matrix} -4v\dot{u}(x-h)-2(x-h/2) \\ +4t(u+u(x-h))-16t^2 \end{matrix} \\ 2(u+u(x+h))+8t & -2\left(1-h^2z\right)(v-u(x-h))/h \end{pmatrix},$$

as in the case of Painleve I. Just as in the continuous situation, the discrete Painleve test can also have the requisite group of invariance. In the continuous case (see Section 3.2), the Painleve property is defined as the absence of the critical movable singularities of an equation,

$$F\left(x, u, x_x, ..., u_{Nx}\right) = 0 \ .$$

A singularity is said to be movable if its position depends on the initial condition and is critical if the solution is multivalued around it. The Painleve expansion is invariant under a group of birotional transformation,

$$(u, x) \rightarrow (V, X)$$
$$u = \gamma\left(x, V, V_x, ..., B_{N-1x}\right) = 0 \qquad\qquad (4.1.45)$$
$$x = P(X) \ ,$$

or in the reverse way

$$(V, X) \rightarrow (u, x)$$
$$U = R\left(X, u, u_X, ..., u_{N-1x}\right) = 0 \qquad\qquad (4.1.46)$$
$$X = Q(x) \ .$$

A simple example is

$$(u,x) \rightarrow (V,X) : u = aV + b/cV + d ; \quad X = Q(x) , \quad ad - bc \neq 0 .$$

The discrete counter part is

$$E\big(x, h\{u(x+kh), k - k_0 = 0, ..., N\}\big) = 0$$

$$E\left(x, q\{u\big(xq^k\big), k - k_0 = 0, ..., N\}\right) = 0 \qquad (4.1.47)$$

for all x and h.

We then state that a discrete equation is said to have a discrete Painleve property if a neighborhood of $h = 0$ or $q = 1$ at every point of which the general solution $x \rightarrow u(x,h)$ has no movable critical singularity. The discrete expansion is invariant under an analogous transformation,

$$u = \gamma(x, h, U, u(x+h), ...)$$

$$u = R(x, H, u, u(x+h), u(x-h), ...)$$

$$X = \xi(x,h), \quad H = \eta(h) \quad \text{etc.}$$

It is important to note that the degree of a polynomial degrees of F in $u(x)$ and $u(x + Nh)$, where F is assumed to be a polynomial in the $N + 1$ variables $u(x + kh)$. One may also note that a different method of discretization may change the degree. Suppose we have an equation of the form

$$uu'' - \frac{1}{2}u^2 = 0 ,$$

whose solution is

$$u = (C_1 x + C_2)^2 .$$

Its exact discretization is

$$u\left\{\frac{u(x+h) - 2u(x) + u(x-h)}{h^2}\right\} - 1/2 \frac{(u(x+h) - u(x))}{h^2}$$

$$\times (u(x) - u(x-h)) - h^2/8\left(\frac{u(x+h) - 2u + u(x-h)}{h^2}\right)^2 = 0 . \qquad (4.1.48)$$

Although this procedure does not conserve the degree, there is another one that keeps it intact. It is obtained by dropping the last term

$$u\frac{u(x+h)-2u+u(x-h)}{h^2} - \frac{1}{2}\frac{(u(x+h)-u)(-u(x-h))}{h^2} = 0 \ . \tag{4.1.49}$$

It is possible to linearize both by the following substitutions,

$$u = \psi^2$$

$$u = \frac{1}{4}\left[\psi(\psi(x+h)+\psi(x-h))+\psi^2+\psi(x+h)\times\psi(x-h)\right]$$

to the same form

$$\psi(x+h) - 2\psi(x) + \psi(x-h) = 0 \ .$$

The problem of Painleve analysis for the discrete equation is very difficult due to the arithmetics involved in it, but one can always use the notion of limits to convert it into a problem of analysis. Consider an arbitrary discrete problem that depends on some parameter a, and let $(x,h,u,a) \rightarrow (X,H,U,A,\varepsilon)$ be an arbitrary possible perturbation in conformity with the theorem of Poincare. We start by discretizing equations (4.1.48) and (4.1.49).

$$F = -(u(x+h)-2u+u(x-h))h^{-2} + 3\lambda_1(u(x+h)+u(x-h))u$$

$$+6\lambda_2 u^2 + 6\lambda_3 u(x-h)u(x+h) + g = 0$$

$$F = -(u(x+h)-2u+u(x-h))h^{-2} + \lambda_1(u(x+h)+v(x-h))$$

$$\times u^2 + 2\lambda_2 u^3 + 2\lambda_3 u(x+h)(u(x))u(x-h)$$

$$+x\left[\mu_1(u(x+h)+u(x-h))/2 + \mu_2 u\right] + \alpha = 0 \ ,$$

with $\sum \lambda_k = 1$, $\sum \mu_k = 1$, α a constant and g an unspecified function of x. The test will impose conditions on (λ_i, μ_i, g). It is important to consider the following cases:

(a) $PI: g = x, \ \lambda = (2/3, 1/3, 0)$

(b) $PII: \lambda = (1, 0, 0), \ \mu = (0, 1),$

for which there exist Lax pairs.

The method proposed by Conte *et al.* is a perturbation of the continuum limit entirely analogous to the Fuchsian or non-Fuchsian perturbative method.

$$x - \text{ unchanged}, \quad h = \varepsilon, \quad q = e^{\varepsilon}, \quad u = \sum_{n=0}^{\infty} \varepsilon^n u^{(n)}$$

$$a = \text{analytic } (A, \varepsilon),$$

which leads to an infinite sequence of differential equations $F^{(n)} = 0$ defined by

$$F = \sum \varepsilon^n F^{(n)}$$

$$F^{(n)}\left(x, u^0, \ldots, u^n\right)$$

$$= F^0\left(x, u^0\right)' u^n + R^n\left(x, u^0, \ldots, u^{n-1}\right) = 0, \quad n > 1,$$

where the first $n = 0$ is the "continuum limit". The next ones $n \geq 1$, which are linear homogeneous, have the same homogeneous part independent of n, defined by the derivative of the equation of the continuum limit, while their inhomogeneous part $R^{(n)}$ comes at the same time from the nonlinearities and discretization. An example by Conte *et al.* considers the Euler scheme for the Bernoulli equation

$$F = \{(u(x+h) - u(x))\} / h + u^2 = 0,$$

which is nothing but the logistic map of Verhulst, known to exhibit chaotic behavior. If we expand the terms in the above equation up to an order ε, sufficient to build the equation $F^1 = 0$, beyond the continuum limit $F^0 = 0$

$$u = u^{(0)} + u^{(1)}\varepsilon + \ldots$$

$$u^2 = u^{(0)^2} + 2\left(u^{(0)}u^{(1)}\varepsilon + \ldots\right)$$

$$\bar{u} = u + uh + \frac{1}{2}uh^2 + \ldots$$

$$\frac{u(x+h) - u(x)}{h} = u^{(0)'} + \left(u^{(1)'} + \frac{1}{2}u^{(0)''}\right)\varepsilon + \ldots.$$

The equations of orders $n = 0$, $n = 1$ are

$$F^0 = u^{(0)'} + u^{(0)^2} = 0$$

$$F' = E^{(0)'} u^{(1)} + \frac{1}{2} u^{(0)''} = 0$$

$$E^{(0)'} = \partial_x + 2u^{(0)} \ .$$

The general solution is

$$u^{(0)} = \chi^{-1}, \quad \chi = x - x_0$$

$$u^{(1)} = u_{-1}^{(1)} \chi^{-2} - \chi^2 \log \psi$$

$$\psi = x - x_0$$

$$u_{-1}^{(1)} = x - \text{arbitrary} \ ,$$

and the movable logarithm proves the instability at $n = 1$.

The subject of discrete Painleve equation is more complicated than the continuous one and it needs more analysis and study. We do not go into further details. Refer to the original literature for more information.

4.2 Long wave/short wave equation

A simple nonlinear system that has received wide attention is the equation describing the interaction of a long and short wave in a fluid. Such a system can be written as

$$A_t = 2S(BB^*)_x$$

$$B_t - 2iB_{xx} = K_2 A_x B - K_3 AB_x + iK_4 A^2 B - 2iSB^2 B \ , \tag{4.2.1}$$

where K_2, K_3, K_4 and S are arbitrary constants of which K_2 and K_3 may be complex. A Painleve analysis of such a coupled system was performed by Roy Chowdhury and Chanda[43,48] in two publications. The first one treated special values of K_2, K_3 and K_4, and the second attempted to proceed with general values of these parameters, with constraints deduced from the Painleve analysis. It was observed that the analysis could single out the important situations originally studied by Newell for which he was able to write down Lax pair, whereas the several other cases could not conform to the criterion of complete integrability. The form of equation discussed by Newell is related to equation (4.2.1) by a scaling transformation.

Leading order analysis

We can rewrite equation (4.2.1) using the complex conjugate of the second one in the following way:

$$A_t = 2S(BC)_x$$

$$B_t - 2iB_{xx} - K_2 A_x B - K_3 AB_x + iK_4 A^2 B - 2iSB^2 C$$

$$C_t + 2iC_{xx} = K_2 A_x C - K_3 AC_x - iK_4 A^2 C + 2iSC^2 B \ . \tag{4.2.2}$$

As before, we set

$$A = \phi^\alpha \sum_{j=0}^\infty a_j \phi^j \ ; \quad B = \phi^\beta \sum_{j=0}^\infty b_j \phi^j \tag{4.2.2a}$$

$$C = \phi^\gamma \sum_{j=0}^\infty C_j \phi^j \ .$$

For the leading order analysis, we assume

$$A \sim a_0 \phi^\alpha \ , \quad B \sim b_0 \phi^\beta \ , \quad C \sim C_0 \phi^\gamma$$

$\phi = x - f(t)$. One can then deduce that

$$\alpha = -1$$

$$\beta + \gamma = -1 \tag{4.2.3}$$

$$a_0 \dot{f} = -2Sb_0 C_0$$

$$-2i\beta(\beta - 1) = -(K_2 + K_3\beta)a_0 + iK_4 a_0^2$$

$$+2i\gamma(\gamma - 1) = -(\overline{K}_2 + \overline{K}_3\gamma)a_0 - iK_4 a_0^2 \ ,$$

corresponding to the dominant terms

$$A_t = 2S(BC)_x$$
$$-2iB_{xx} = K_2 A_x B - K_3 AB_x + iK_4 A^2 B$$
$$+2iC_{xx} = \overline{K}_2 A_x C - \overline{K}_3 AC_x - iK_4 A^2 C .$$

Here $\overline{K}_2$ and $\overline{K}_3$ denote the complex conjugate of K_2 and K_3. Equation (4.2.3) indicates that β and γ, are not uniquely determined, so we set

$$\beta = -p \quad (p \text{ real}),$$

then $\gamma = p-1$, along with

$$a_0 \dot{f} = -2Sb_0 C_0$$
$$-2ip(p+1) = -(K_2 - K_3 p)a_0 + iK_4 a_0^2 \tag{4.2.4}$$
$$2i(p-1)(p-2) = -\left[\overline{K}_2 + \overline{K}_3(p-1)\right]a_0 - iK_4 a_0^2 .$$

Here we have two equations quadratic in a_0 and p. To be consistent, we must have

(a) either both the roots for a_0 are the same,

$$\frac{-p(p+1)}{(p-1)(p-2)} = \frac{K_2 - K_3 p}{\overline{K}_2 + \overline{K}_3(p-1)} = -1 , \tag{4.2.5}$$

which leads to

$$p = 1/2 \quad \text{and} \quad 2(K_2 + \overline{K}_2) = (K_3 + \overline{K}_3) . \tag{4.2.6}$$

We also get a single equation for a_0,

$$2iK_4 a_0^2 - (2K_2 - K_3)a_0 + 3i = 0 .$$

Note that the case $K_4 = 0$, $p = 1/2$ is included in this case with

$$a_0 = \frac{6i}{2(K_2) - K_3} ;$$

(b) or when the roots are different, then

$$K_2 + \overline{K}_2 + K_3(p-1) - K_3 p \neq 0 \ ,$$

whence we get

$$a_0 = \frac{4i(2p-1)}{K_2 + \overline{K}_2 + \overline{K}_3(p-1) - K_3 p} \ . \tag{4.2.7}$$

We can also solve for a_0^2 and get

$$a_0^2 = \frac{-2\{p(p+1)[\overline{K}_2 + \overline{K}_3(p-1)] + [(p-1)(p-2)(K_3 - K_3 p)]\}}{K_4[K_2 + \overline{K}_2 + (p-1)\overline{K}_3 - K_3 p]} \ . \tag{4.2.8}$$

Equations (4.2.7) and (4.2.8) imply that

$$8(2p-1)^2 K_4 = \{p(p+1)[\overline{K}_2 + \overline{K}_3(p-1)]$$
$$+(p-1)(p-2)(K_2 - K_3 p)\}[\overline{K}_2 + \overline{K}_3(p-q) - K_3 p] \ . \tag{4.2.9a}$$

It may be noted that $p = 1/2$ is not allowed in equation (4.2.7). In the special case, $K_2 = -2$, $K_3 = 0$, $K_4 = 2$, $\overline{K}_2 = -2$, $\overline{K}_3 = 0$, we get from (4.2.7) $p = 0, 1$. Therefore, we have two distinct situations:

(i) $p = 0$, $a_0 = i$,

(ii) $p = 1$, $a_0 = -i$.

Recursion relation

In the general situation, we can now set

$$A = \sum a_j \phi^{j-1} \ , \quad B = \sum b_j \phi^{j-p} \ , \quad C = \sum c_j \phi^{j+p-1} \ ,$$

with a_j, b_j, c_j function of t. When substituted in equation (4.2.2) we get

$$\begin{pmatrix} P & Q & 2sb_0(n-1) \\ R & S & 0 \\ T & 0 & \Lambda \end{pmatrix} \cdot \begin{pmatrix} a_n \\ b_n \\ c_n \end{pmatrix} = \begin{pmatrix} X \\ Y \\ Z \end{pmatrix} \qquad (4.2.9b)$$

$$P = (n-1)\dot{f}$$

$$R = \left[K_2(n-1) + pK_3 + 2iK_4 a_0 \right] b_0$$

$$T = \left[\overline{K}_2(n-1) - (p-1)\overline{K}_3 - 2iK_4 a_0 \right] c_0$$

$$Q = 2SC_0(n-1)$$

$$S = 2i(n-1)(n-p-1) - K_2 a_0 + iK_4 a_0^2 - K_3 a_0(n-p)$$

$$\Lambda = -2i(n+p-1)(n+p-2) - \overline{K}_2 a_0 - iK_4 a_0^2 - \overline{K}_3 a_0(n+p-1),$$

whereas the entries in the vector on the right side of equation (4.2.9) are given as

$$X = \dot{a}_{n-1} - 2S \sum_{q=1}^{n-1} b_{n-q} c_q (n-1)$$

$$Y = \dot{b}_{n-2} - b_{n-1}(n-p-1)\dot{f} - K_2 \sum_{q=1}^{n-1} a_{n-q} b_q (n-q-k)$$

$$+ K_3 \sum_{q=1}^{n-1} a_{n-q} b_q (q-p) - iK_4 \sum_{\substack{d=0 \\ d+q>0}}^{n-1} \sum_{q=0}^{n-1} a_{n-q-d} a_d b_q$$

$$+ 2iS \sum_{d=0}^{n-1} \sum_{q=0}^{n-1} b_{n-q-d-1} c_d b_q$$

$$Z = \dot{c}_{n-2} - c_{n-1}(n+p-2)\dot{f} - \overline{K}_2 \sum_{q=1}^{n-1} a_{n-q} c_q (n-q-1)$$

$$+ \overline{K}_3 \sum_{q=1}^{n-1} a_{n-q} c_q (q+p-1) + iK_4 \sum_{\substack{d=0 \\ d+q>0}}^{n-1} \sum_{q=0}^{n-1} a_{n-q-d} a_d c_q$$

$$- 2iS \sum_{d=0}^{n-1} \sum_{q=0}^{n-1} c_{n-q-d-1} c_d b_q . \qquad (4.2.10)$$

The resonance positions are now determined by the vanishing of the system matrix occurring on the left side of equation (4.2.9). It turns out to be

$$4\dot{f}(\gamma+1)\gamma(\gamma-1)(\gamma-\sigma)(\gamma-\tau)=0\ ,\qquad\qquad\qquad (4.2.11)$$

so the resonances are

$$\gamma=-1,0,1,\sigma,\tau\ ,$$

where σ,τ are roots of

$$4\gamma^2+\sigma'\gamma+\sigma''=0\ ,$$

where σ',σ'' are given as

$$\sigma'=1/4\left(-20+2iK_3a_0-2i\overline{K}_3a_0+2ia_0\overline{K}_2-2iK_2a_0\right)$$

$$\sigma''=1/4\Big[\left(-32p^2+32p+16\right)+(6p-8)iK_3a_0$$
$$-(6p+2)\left(i\overline{K}_3a_0\right)-(4p-6)iK_2a_0-(4p+2)$$
$$\times iK_2a_0+a_0^2\left(K_3\overline{K}_3-K_2\overline{K}_3-K_3\overline{K}_2\right)\Big]\ .$$

Now $\gamma=-1$, correspond to the arbitrariness of $\phi=x-f(t)$, where $\gamma=0$ to that of either b_0 or c_0. But for other resonance positions to be meaningful it must occur at integer point. Therefore, we must have

$$\begin{aligned}\sigma+\tau &= \text{integer }(+Ve)\\[4pt]\sigma\tau &= \text{integer }(+Ve)\end{aligned}\qquad\qquad (4.2.12)$$

However, when σ, and τ are negative, the expansions of the functions A, B, C may possibly represent a special solution in the neighborhood of $\phi(x,t)=0$. From equation (4.2.7) we observe that

$$a_0=\frac{2i[(\sigma+\tau)-5]}{\left(K_3-\overline{K}_3\right)-\left(K_2-\overline{K}_2\right)}\ .$$

For a_0 to be nonzero, we must simultaneously have

$$\sigma + \tau = 5$$
$$K_3 - \overline{K}_3 = K_2 - \overline{K}_2 \ .$$

For other combinations, we must have $\sigma + \tau \neq 5$.

Now equations (4.2.11) and (4.2.12) imply that

$$\sigma = 3 + i/4\left[2\left(K_2 - \overline{K}_2\right) - \left(K_3 - \overline{K}_3\right)\right]a_0$$

$$\tau = 2 - i/4\left(K_3 - \overline{K}_3\right)a_0 \ ,$$

(4.2.13)

from which we deduce

$$i\left(K_3 - \overline{K}_3\right)a_0 = 4(2 - \tau)$$
$$i\left(K_2 - \overline{K}_2\right)a_0 = 2(\sigma - \tau) - 2 \ .$$

These, along with equation (4.2.11), yield the following choices:

(a) $\tau = 2, \ \sigma = 3$ or $\sigma = 3, \ \tau = 2$

For $\tau = 2, \ \sigma = 3$ we get

$$K_3 = \overline{K}_3 , \quad K_2 = \overline{K}_2 \ .$$

For $\tau = 3, \ \sigma = 2$ we get

$$a_0 = 4i/\left(K_2 - \overline{K}_2\right) \ ,$$

along with $K_3 - \overline{K}_3 = K_2 - \overline{K}_2$, but $K_3 \neq \overline{K}_3 K_2 \neq \overline{K}_2$. Therefore, from equations (4.2.11), (4.2.12), and (4.2.13), we are led to

$$\overline{K}_3 = 1/2\left(K_2 + 3\overline{K}_2\right)$$

$$K_3 = 1/2\left(3K_2 + \overline{K}_2\right)$$

$$K_4 = 1/32\left(K_2 - \overline{K}_2\right)^2 \ .$$

(b) $\sigma = 5,\ \ \tau = 0$ or $\sigma = 0,\ \ \tau = 5$

   $\sigma = 5,\ \ \tau = 0$ leads to,

$$K_3 = 1/2\big(3K_2 + \overline{K}_2\big),\qquad \overline{K}_3 = 1/2\big(K_2 + 3\overline{K}_2\big)$$

$$a_0 = -8i/\big(K_2 - \overline{K}_2\big),\qquad K_4 = 7/128\big(K_2 - \overline{K}_2\big)^2,$$

whereas $\tau = 5$ and $\sigma = 0$ yield the same value of K_3 and $\overline{K}_3$, but

$$a_0 = 12i/\big(K_3 - \overline{K}_3\big),\qquad K_4 = -3/288\cdot\big(K_2 - \overline{K}_2\big)^2.$$

(c) $\sigma = 4,\ \ \tau = 1$

$$K_4 = -(5/32)\cdot\big(K_2 - \overline{K}_2\big)^2,\qquad K_3 = 1/2\cdot\big(3K_2 + \overline{K}_2\big)$$

$$K_3 = 1/2\cdot\big(3\overline{K}_2 + K_2\big),\qquad a_0 = -4i/\big(K_3 - \overline{K}_3\big).$$

(d) $\tau = 4,\ \ \sigma = 1$

$$K_4 = 1/128\big(K_2 - \overline{K}_2\big)^2,\qquad K_3 = 1/2\big(3K_2 + \overline{K}_2\big)$$

$$K_3 = 1/2\big(3\overline{K}_2 + K_2\big),\qquad a_0 = 8i/\big(K_3 - \overline{K}_3\big).$$

(e) Now consider situations other than those enumerated above. Here we get

$$K_3 - \overline{K}_3 = \big(K_2 - \overline{K}_2\big)\lambda,\qquad \lambda = 2(2-\tau)/(\sigma - \tau - 1)$$

$$\overline{K}_3 = 1/2\big[(\lambda+2)\overline{K}_2 - (\lambda-2)K_2\big]$$

$$K_4 = 1/\big[16(2-\tau)^2\big]\cdot\big[3/2\lambda^2 - (2-\tau)\lambda(\lambda-2)\big]\big(K_2 - \overline{K}_2\big)^2$$

$$K_3 = 1/2\big[(\lambda+2)K_2 - (\lambda-2)\overline{K}_2\big],\qquad \tau \neq 2.$$

However, another series of cases originates from equations (4.2.7) and (4.2.8). When

$$K_3 - \overline{K}_3 = K_2 - \overline{K}_2$$

along with $\sigma + \tau = 5$, we have

$$a_0 = 4i(2p-1)/(K_2 + \overline{K}_2 + \overline{K}_3(p-1) - K_3 p)$$

$$8(2p-1)^2 K_4 = \left\{ p(p+1) \left[\overline{K}_2 + \overline{K}_3(p-1) \right] + (p-1)(p-2)(K_2 - K_3 p) \right\} \left[K_2 + \overline{K}_2 \right.$$

$$+ \overline{K}_3(p-1) - K_3 p \right] \left[\left(-32p^2 + 32p + 16 \right) + (6p-8)iK_3 a_0 + (6p+2)(iK_3 a_0) \right.$$

$$\left. -(4p-6)iK_2 a_0 - (4p+2)i\overline{K}_2 a_0 + a_0^2 \left(K_3 \overline{K}_3 - K_2 \overline{K}_3 - K_3 \overline{K}_2 \right) \right] 4\sigma(5-\sigma). \quad (4.2.14a)$$

If we eliminate a_0, we get two fourth-degree equations in p. The consistency of the roots will give conditions for the resonance positions at integer γ values. This situation leads to the case considered by Newell, that is $p = 0$, $a_0 = i$, and $p = 1$, $a_0 = -i$. In these cases, we get the resonance positions to be, $-1, 0, 1, 2, 3$.

Another situation arises when equation (4.2.12) does not hold. Here we have

$$a_0 = 2i[(\sigma + \tau) - 5]/\left(K_3 - \overline{K}_3 - \left(K_2 - \overline{K}_2 \right) \right) \qquad (4.2.14b)$$

and

$$a_0 = 4i(2p-1)/\left(K_2 + \overline{K}_2 + \overline{K}_3(p-1) - K_3 p \right) . \qquad (4.2.15)$$

Equating these two expressions at once yields

$$p = \frac{2\left[\left(K_3 - \overline{K}_3 \right) - \left(K_2 - \overline{K}_2 \right) \right] + \left(K_2 + \overline{K}_2 - \overline{K}_3 \right)(\sigma + \tau - 5)}{4\left[\left(K_3 - \overline{K}_3 \right) - \left(K_2 - \overline{K}_2 \right) \right] + \left(K_3 - \overline{K}_3 \right)(\sigma + \tau - 5)} ,$$

along with

$$8(2p-1)^2 K_4 = \left\{ p(p+1) \left[\overline{K}_2 + \overline{K}_3(p-1) \right] + (p-1)(p-2) \right.$$

$$\times (K_2 - K_3 p) \right\} \left[K_2 + \overline{K}_2 + \overline{K}_3(p-1) - K_3 p \right] \qquad (4.2.16a)$$

and equation (4.2.14a) with the right side replaced by $4\sigma\tau$.

Search for arbitrary coefficients
Go back to equation (σ) and check that if

$$\left[K_2(\gamma - 1) + pK_3 + 2iK_4a_0\right] = M_1$$

$$\left[K_2(\gamma - 1) - (p-1)K_3 - 2i\left(K_4a_0\right)\right] = M_2$$

$$\left[-2i(\gamma - p)(\gamma - p - 1) - K_2a_0 - K_3a_0(\gamma - p) + iK_4a_0^2\right] = M_3$$
(4.2.16b)

$$\left[-2i(\gamma + p - 1)(\gamma + p - 2) - \overline{K}_2a_0 - \overline{K}_3a_0(\gamma + p - 1) - iK_4a_0^2\right] = M_4 \ ,$$

then the compatibility conditions can be written as follows:

(i)   When $M_1 \neq 0$, $M_2 \neq 0$, $M_3 \neq 0$, $M_4 \neq 0$, then

$$M_3M_4X - 2Sc_0(\gamma - 1)M_4Y - 2Sb_0(\gamma - 1)M_3Z = 0 \qquad (4.2.17)$$

with $n = r$.

(ii)   When $M_1 = 0$, $M_2 \neq 0$, $M_3 = 0$, $M_4 \neq 0$, we get

$$Y = 0$$

with $n = \gamma$.

(iii) When $M_1 \neq 0$, $M_2 = 0$, $M_3 \neq 0$, $M_4 = 0$, we have

$$Z = 0$$

with $n = \gamma$.

Now consider some particular cases:

Case I.
$$K_2 = -2, \ K_3 = 0, \ K_4 = 2, \ \overline{K}_2 = -2, \ \overline{K}_3 = 0$$

In this case we have two branches:

$$p = 0, \ a_0 = i \quad \text{and} \quad p = 1, \ a_0 = -i \ .$$

For both, the branches of the resonance positions are $-1, 0, 1, 2, 3$. The positions $\gamma = -1$, and 0 correspond, respectively, to the arbitrariness of $\phi(x,t)$ and (b_0, C_0). For the other positions $(p = 0, a = i)$, we have the following:

At $\gamma = 1$:

$$a_1 = -\frac{1}{4}\dot{f}$$
$$b_1 = \text{arbitrary} \qquad\qquad\qquad (4.2.18)$$
$$c_1 = -i\,c_0\dot{f}/4 \ ,$$

along with a condition $\dot{a}_0 = 0$, which is obtained because the first row of the matrix (4.2.9) vanishes. One can observe that this condition is satisfied identically.

At $\gamma = 2$: Here the condition (i) above is satisfied and therefore the compatibility condition is given as in equation (4.2.17) with particular values of $K_1, \overline{K}_2, p, a_0, \gamma$. The expansion coefficients are given as

$$a_2 = \text{arbitrary}$$
$$b_2 = -\frac{i}{32}\left(8b_0 - 3i\dot{f}^2 b_0 + 48b_0 a_0\right) \qquad\qquad (4.2.19)$$
$$C_2 = -\frac{i}{32}\left(8C_0 + 3i\dot{f}^2 C_0 - 16C_0 a_2 - 16iSC_0^2 b_1\right) \ .$$

At $\gamma = 3$: Here (iii) is satisfied, and the coefficients are

$$a_3 = \frac{1}{2\dot{f}}\left(-4SC_0 b_3 + \dot{a}_2 - 4Sb_0 C_3 - 4S[b_1 C_2 + b_2 C_1]\right)$$
$$b_3 = \frac{1}{12i}\left(8b_0 a_3 + \dot{b}_1 + 6b_1 a_2 - 4ia_1 b_0 a_2 - 4a_1 b_2 + 2iSb_0^2 C_2\right.$$
$$\left. -2ia_1^2 b_1 + 2iSb_1^2 C_0 + 4iSb_0 b_1 C\right)$$
$$C_3 = \text{arbitrary}.$$

Similarly for the branch $p = 1$, $a_0 = -i$, we have

$\gamma = 1$:
$$a_1 = -\dot{f}/4$$
$$b_1 = ib_0\dot{f}/4$$
$$C_1 = \text{arbitrary} \quad \text{and} \quad \dot{a}_0 = 0$$

$\gamma = 2$:

$\quad a_2 = $ arbitrary

$$b_2 = \frac{i}{32}\left(8\dot{b}_0 - 3if\dot{C}_0 - 16b_0a_2 + 16iSb_0^2C_1\right)$$

$$C_2 = \frac{i}{32}\left(8\dot{C}_0 + 3if^2C_0 + 48C_0a_2\right)$$

$\gamma = 3$:

$$a_3 = 1/(2\dot{f})\cdot\left(-4SC_0b_3 - 4Sb_0C_3 + \dot{a}_2 - 4Sb_1C_2 - 4Sb_2C_1\right)$$

$\quad b_3 = $ arbitrary

$$C_3 = -1/(12i)\cdot\left(8C_0a_3 + \dot{C}_1 + (6C_1 + 4ia_1C_0)a_2\right.$$
$$\left. +4a_1C_2 - 2iSC_0^2b_2 + 2ia_1^2C_1 + 2iSC_1^2b_0 - 4SC_0b_1C_1\right).$$

Thus, for both the branches (4.2.2a) represents a general solution containing a maximum number of arbitrary functions allowed at the resonances satisfying the Cauchy-Kovalevsky theorem, and it is possible to infer that the system is completely integrable.

Case II.

$$p = 1/2, \; K_2 = \overline{K}_2, \; K_3 = \overline{K}_3, \; 2K_2 = K_3$$
$$\gamma = -1, 0, 1, 2, 3.$$

This is a special situation with $\tau = 2$, $\sigma = 3$ of the general case discussed previously.

(a) $\gamma = -1$ corresponds to arbitrary $\phi = x - f(t)$

(b) $\gamma = 0$ corresponds to the arbitrariness of the coefficients b_0 or C_0.

(c) $\gamma = +1$, therefore we get

$\quad a_1 = $ arbitrary

$$b_1 = \frac{-\left(\dot{f} - 2ia_0\dot{f}\right)b_0 + (K_3 + 4iK_4a_0)b_0a_1}{2(2i + K_3a_0)}$$

$$C_1 = \frac{-\left(\dot{f} + 2ia_0\dot{f}\right)C_0 + (K_3 - 4iK_4a_0)C_0a_1}{2(-2i + K_3a_0)}$$

(d) $\gamma = +2$.

Here we get the following relation to be satisfied, which is a differential equation for a_1:

$$\dot{a}_1 - 2Sb_1C_1 + \frac{SC_0}{K_3a_0}\left\{\dot{b}_0 - \frac{b_1\dot{f}}{2} + \frac{K_3a_1b_1}{2} - 2iK_4a_0a_1b_1\right.$$

$$+4iSb_1b_0C_0 + 2iSb_0^2C_1\right\} + \frac{Sb_0}{K_3a_0}\left\{\dot{C}_0 - \frac{C_1\dot{f}}{2} + \frac{K_3a_1b_1}{2}\right.$$

$$+2iK_4a_0a_1C_1 - 4iSC_1C_0b_0 - 2iSC_0^2b_1\right\} = 0$$

a_1 is fixed, and we cannot meet the requirement of the Cauchy-Kovalevsky theorem. So in this situation, the equation does not pass the test.

Case III.
$$p = 1/2, \quad K_4 = 1/32\left(K_2 - \overline{K}_2\right)^2, \quad K_3 \neq \overline{K}_3$$
$$K_2 \neq \overline{K}_2, \quad \overline{K}_3 = 1/3\left(3\overline{K}_2 + K_2\right)$$
$$K_3 = 1/2\left(3K_2 + \overline{K}_2\right).$$

Here we get $a_0 = 8i/\left(K_3 - \overline{K}_3\right)$, and a_1 becomes determined due to the compatibility condition at $\gamma = 2$, so this case is also not completely integrable.

Case IV.
$$p = 1/2, \quad K_3 = 1/2\left(3K_2 + \overline{K}_2\right), \quad K_3 \neq \overline{K}_3$$
$$K_2 \neq \overline{K}_2, \quad \overline{K}_3 = 1/2\left(K_2 + 3\overline{K}_2\right), \quad K_4 = 7/128\left(K_2 - \overline{K}_2\right)^2.$$

Here we have $a_0 = -8i/\left(K_3 - \overline{K}_3\right)$. This corresponds to the special values $\sigma = 0, \ \tau = 5$ of the general case. The resonances are found to be at $\gamma = -1, 0, 0, 1, 5$. In this situation, a_0 is also fixed and only one of the coefficients b_0 and C_0 can be arbitrary. So for the same reason as above, this case also does not pass the test.

In a similar manner one can analyze all the situations given after equation (4.2.13). Since the analysis is straightforward, we omit their discussions.

The case of long wave/short wave equation shows how the various coefficients occurring in a given nonlinear partial differential equation affect the various features of the Painleve test.

4.3 Fermionic system and Painleve analysis

A different class of nPDEs arose when nonlinearity became important in the domain of elementary particles. Because elementary particles are basically of two types – bosons and fermions – there are two kinds of fields: commuting and anticommuting. A different and unified approach was later put forward by Salam and Strathde, and Weiss and Zumino to treat bosons and fermions the same. This is known as the supersymmetric approach. People have already formulated supersymmetric versions of Sine-Gordon, nonlinear Schrodinger, KdV equation, and many other systems. Some of them have turned out to be completely integrable, but not all. But there are more than one super versions of the same nonlinear equation, such as the KdV equation. There are already three different forms of the super KdV equation (SKdV) – one due to Kupershmidt, the others due to Mathieu *et al.*[86]

Naturally, the question of complete integrability of such equations requires extensive analysis. Because the equations under consideration are classical, we are to deal with fermionic variables, the use of grassmanian quantities are square. One should remember that square of any grassmanian variable is 0 due to the basic anticommuting nature. Below is a discussion of three examples – super Burger equation, super nonlinear Schrodinger, and, of course, the super KdV equation.

The Super KdV equation

From the beginning of the soliton theory, the KdV equation has been used as the testing ground for almost all the developments of the theory. Hence, it is quite proper to analyze the case of the super KdV problem first. As has already been said, not all versions of SKdV are completely integrable. Also, not all of them are "supersymmetric," which means that not all of them are invariant under supersymmetric transformations. Some are constructed by simply adjoining a fermionic equation with the original one via a guessed form of coupling term. Below is the treatment of Mathieu.[86,87]

The equation being considered is written as

$$u_t = -u_{xxx} + 6uu_x - 3\xi\xi_{xx}$$
$$\xi t = -c\xi_{xxx} + b\xi_x u + a\xi u_x \ , \tag{4.3.1}$$

where $a, b,$ and c are numerical factors that distinguish between the bosonic and fermionic fields to introduce a grading: $\deg x = -1$, $\deg t = -3$, $\deg u = 2$, $\deg \xi = 3/2$. Usually the fermionic variable ξ is assigned a half integral grading. For the Painleve test, we expand both the bosonic and fermionic field variables as

$$u = \sum_{j=0}^{\infty} u_j \phi^{\alpha+j} , \quad \xi = \sum_{j=0}^{\infty} \xi_j \phi^{\beta+j} .$$

For the leading singularities $u \sim u_0 \phi^{\alpha}$ and $\xi \sim \xi_0 \phi^{\beta}$, we get $\alpha = \beta = -2$. The singularity manifold ϕ is assumed to be of the form $\phi(x,t) = x - f(t)$. If we further assume that u_j, ξ_j are functions of t, substituting the series for u and ξ in equation (4.3.1) gives us

$$u_{j-3,t} + (j-4)u_{j-2}/t + (j-2)(j-3)(j-4)u_j - 3(j-4)\sum_{k=0}^{j} u_k u_{j-k}$$

$$+3/2(j-4)\sum_{k=0}^{j+1} \xi_k \xi_{j+1-k}(j+1-2k) = 0$$

$$\xi_{j-3,t} + (j-4)\xi_{j-2},\phi_t + C(j-2)(j-3)(j-4)\xi_{j-b}\sum_{k=0}^{j} u_k \xi_{j-k}(j-k-2)$$

$$-a\sum_{k=0}^{j} \xi_k u_{j-k}(j-k-2) = 0 .$$

Proceeding as before, we can read off the determinant whose roots will determine the resonance position. This is

$$\begin{vmatrix} (j-2)(j-3)(j-4) - 6u(j-4) & -3\xi_0\xi_1(j-4)(j-3) \\ 2a+2b-a_j & C(j-2)(j-3)(j-4) + u_0\left(2a+2b-a_j\right) \end{vmatrix} . \quad (4.3.2)$$

Here

$$u_0 = 2 + \frac{1}{2}\xi_0\xi_1 ,$$

then the vanishing of the above determinant leads to

$$[(j-2)(j-3)(j-4) - 12(j-4)][C(j-2)(j-3)(j-4) + 2(2a+2b-bj)]$$

$$+\xi_0\xi_1 p(j) = 0 . \quad (4.3.3)$$

$P(j)$ is of fourth degree in j. Assume that $\xi_0\xi_1$ is arbitrary. Then the two terms in equation (4.3.3) should vanish separately. One such condition is

$$(j+1)(j-4)(j-6)\left[C(j-2)(j-3)(j-4) + 2\left(2a+2b-b_j\right)\right] = 0 , \quad (4.3.4)$$

whence we get three resonances at $j = -1, 4, 6$. On the other hand, three more roots should come from the terms within the square bracket in equation (4.3.4). If the three roots are j_1, j_2, and j_3, then

$$j_1 + j_2 + j_3 = 9$$
$$j_1 j_2 + j_2 j_3 + j_3 j_1 = 26 - 2b / C \qquad (4.3.5)$$
$$j_1 j_2 j_3 = 24 - 4 / C(b + a) .$$

For the effectiveness of these resonances they must be integers and different. Assume $j_3 > j_2 > j_1$. It can then be observed that we have six solutions:

(1) $j_1 = 0$, $j_2 = 2$, $j_3 = 7 \rightarrow b = 6c$, $a = 0$

(2) $j_1 = 0$, $j_2 = 3$, $j_3 = 6 \rightarrow b = 2a = 4C$

(3) $j_1 = 0$, $j_2 = 4$, $j_3 = 5 \rightarrow b = a = 3C$ $\qquad$ (4.3.6)

(4) $j_1 = 1$, $j_2 = 2$, $j_3 = 6 \rightarrow b = 3C$, $a = 0$

(5) $j_1 = 1$, $j_2 = 3$, $j_3 = 5 \rightarrow b = 2a = 3C / 2$

(6) $j_1 = 2$, $j_2 = 3$, $j_3 = 4 \rightarrow b = a = 0$.

The next point is the compatibility of the coefficients at these positions. The analysis is again straightforward, and it is found that in cases (1), (2), (4), and (6), the system does not pass the test. But for $C = 1$ and $C = 4$, the cases (3) and (5) do satisfy the Painleve criterion. Therefore we have two choices of the coefficients:

(1) $a = b = 3$, $C = 1$
(2) $a = 3$, $b = 6$, $C = 4$.

The first case was considered by Mathieu and is invariant under the super transformation $\delta u = \eta \xi_x$, $\delta \xi = \eta u$, with n a constant anticommuting parameter. The second case was first obtained by Kupershmidt and has been found to be bi-Hamiltonian.

Supersymmetric nonlinear Schrodinger equation
Another important example of supersymmetric nonlinear system is the Super Nonlinear Schrodinger Equation (SNLS), which has received wide attention. Analysis of such nonlinear supersymmetric system was done by Roy Chowdhury and Naskar,[87] where the conserved quantities, recursion operator, and simplistic form were deduced. Complete

integrability via the Painleve test was done by Roy Chowdhury and Naskar. Here we will follow the treatment of reference (87) to show how the Painleve test is applied in this situation.

The SNLS system is written as

$$iq_t = -q_{xx} + 2Kq^*q^2 + K\psi^{+}\psi q - iK^{1/2}\psi\psi_x$$
$$i\psi_t = -2\psi_{xx} + Kq^{+}q\psi - iK^{1/2}\left(2q\psi_x^{+} + \psi^{+}q_x\right),$$

(4.3.7)

where $q(x,t)$ is the original variable satisfying the usual NLS equation in absence of ψ the super counterpart. We set

$$q = u_0 + iv_0$$
$$\psi = u_1 + iv_1,$$

whence we get four coupled equations:

$$u_{0t} = -v_{0xx} + K\left[2v_0\left(u_0^2 + v_0^2\right) + v_0\left(u_1^2 + v_1^2\right)\right]$$
$$\quad -K^{1/2}\left(u_1u_{1x} - v_1v_{11x}\right)$$

$$v_{0t} = +u_{0xx} - k\left[2u_0\left(u_0^2 + v_0^2\right) + u_0\left(u_1^2 + v_1^2\right)\right]$$
$$\quad -k^{1/2}\left(v_1u_{1x} + u_1v_{1x}\right)$$

$$v_{1t} = 2u_{1xx} - Ku_1\left(u_0^2 + v_0^2\right) - K^{1/2}\left[2\left(v_0u_{1x} - u_0v_{1x}\right)\right.$$
$$\quad \left. +\left(u_1v_{0x} - v_1u_{0x}\right)\right]$$

$$u_{1t} = -2v_{1xx} + Kv_1\left(u_0^2 + v_0^2\right) - K^{1/2}\left[2\left(u_0u_{1x} + v_0v_{1x}\right)\right.$$
$$\quad \left. +\left(u_1u_{0x} + v_1v_{0x}\right)\right].$$

(4.3.8)

For the Painleve analysis, we set

$$u_0 = \sum_{j=0}^{\infty} a_j\phi^{j+\alpha}$$

$$v_0 = \sum_{j=0}^{\infty} b_j\phi^{j+\beta}$$

(4.3.9)

$$u_1 = \sum_{j=0}^{\infty} C_j\phi^{j+\delta}$$

But because u_1 and v_1 are fermionic, we must assume that the coefficients C_j, d_j are also whence one should have $C_j^2 = d_j^2 = 0$. The leading order analysis suggests

$$\alpha = \beta = \gamma = \delta = -1 , \qquad\qquad\qquad\qquad\qquad (4.3.10)$$

whereas for the leading coefficients we have

$$-2\phi_x^2 + K\left\{2\left(a_0^2 + b_0^2\right) + C_0^2 + d_0^2\right\} - 2K^{1/2}\left(C_0 d_0/a_0\right)\phi_x = 0$$

$$-2\phi_x^2 + K\left\{2\left(a_0^2 + b_0^2\right) + C_0^2 + d_0^2\right\} + K^{1/2}\left[\left(C_0^2 - d_0^2\right)/b_0\right]\phi_x = 0$$

$$-4\phi_x^2 + K\left(a_0^2 + b_0^2\right)C_0 + 3K^{1/2}\left[\left(a_0 d_0 - b_0 C_0\right)\right]\phi_x = 0 \qquad\qquad (4.3.11)$$

$$-4\phi_x^2 d_0 + K\left(a_0^2 + b_0^2\right)d_0 + 3K^{1/2}\left[\left(a_0 C_0 + b_0 d_0\right)\right]\phi_x = 0 .$$

If we assume that $\phi = x - f(t)$ and set $K = 1$, then from equations (4.3.9) and (4.3.11) we get

$$-4 + a_0^2 + 3a_0 = 0 , \qquad C_0 = d_0$$

or $a_0 = -4$, $a_0 = 1$, when $b_0 = 0$.

On the other hand, equation (4.3.11) admits only $a_0 = 1$. To determine the resonances of a new set,

$$u_0 = a_0\phi^{-1} + a_\gamma \phi^{\gamma-1}$$

$$v_0 = b_0\phi^{-1} + b_\gamma \phi^{\gamma-1}$$

$$u_1 = C_0\phi^{-1} + C_\gamma \phi^{\gamma-1} ,$$

where $C_0 = d_0$, $b_0 = 0$, and $a_0 = 1$. As usual, one assume the coefficients a_i, b_i, etc. are all functions of time only. Then the condition for the non-zero values of a_γ, b_γ, c_γ, and d_γ turns out to be

$$\Delta = \begin{vmatrix} 4-\gamma^2+3\gamma & 0 & C_0\gamma & C_0\gamma \\ 0 & -(\gamma^2-3\gamma) & -C_0(\gamma-2) & C_0(\gamma-2) \\ C_0(\gamma-5) & C_0(\gamma-3) & -(2\gamma^2-\sigma\gamma+3) & -(2\gamma-3) \\ C_0(\gamma-5) & -(\gamma-3)C_0 & -(2\gamma-3) & -(2\gamma^2-\sigma\gamma+3) \end{vmatrix} = 0 \; ,$$

which can be simplified to the form

$$\Delta = \gamma^2(\gamma+1)(\gamma-1)(\gamma-2)(\gamma-3)^2(\gamma-4) = 0 \; .$$

Therefore, the resonances are

$$\gamma = 10, -1, 1, 2, 3, 3, 4 \; .$$

As before, $\gamma = -1$ corresponds to the arbitrariness of ϕ, the singular manifold. Because $\gamma = 0$ is also a resonance, we have the arbitrariness of $C_0 = d_0$, which are not determined. At $\gamma = 11$, we get

$$a_1 = 0, \quad b_1 = -\phi t/2, \quad d_1 = \frac{C_0}{12}\phi_t, \quad C_1 = -\frac{C_0}{12}\phi_t \; .$$

But we know that (ϕ, C_0) are arbitrary, so (b_1, d, C_1) are not determined.

Next at $\gamma = 2$

$$6a_2 + 2(C_2 d_0 + d_2 C_0) = -\left[2(a_1^2 + b_1^2) + 4a_1^2 + 2C_0(C_1 + d_1)\right]$$

$$2b_2 = -\left[4a_1 b_1 + 2C_0 b_1(C_1 + d_1)\right]$$

$$C_0(3a_2 - b_2) + C_2 - d_2 = -d_{0t} - C_0(a_1^2 + b_1^2) - 2a_1 C_1$$

$$C_0(3a_2 + b_2) - C_2 + d_2 = C_{0t} - d_0(a_1^2 + b_1^2) - 2a_1 d_1 \; .$$

But because (b_1, d, C_1) are not determined and the det of the coefficients of (a_2, b_2, C_2, d_2) on the right side is 0, it is observed that they are also arbitrary. A similar conclusion also holds for the situations at $\gamma = 3$ and 4. At least we have the requisite number of resonances and same number of arbitrary coefficients, which conforms to the Cauchy-Kowalevsky condition.

4.4 Algebraic integrability and Painleve analysis

Painleve analysis has served as a practical tool for a long time for the study of properties associated with nonlinear partial and ordinary differential equations. But many points still need to be clarified regarding its connection with the complete integrability of a nonlinear system. One idea about this can be obtained by studying its relation with the existence of conserved quantities, which is another approach for analyzing complete integrability. In this respect, one should mention the idea of algebraic integrability studied independently by Adler *et al.*, Ercolani and Siggia, and Ishui. This chapter introduces the concept and gives a short discussion of algebraic integrability following references (88), (89) and (90). Let us start with the definition.

A Hamiltonian system will be called algebraically completely integrable if it can be linearized on an abelian variety, according to Adler and van Moerebek. On the other hand, Ishui studies the relation between the existence of Laurent expansion and integrable of motion. In both of these approaches, the main intention is to establish the notion of complete integrability in the algebraic way. Adler *et al.* considered the dynamical system with the following Hamiltonian system

$$H = 1/2\sum_{i=1}^{l} p_i^2 + \sum_{i=1}^{l+1} \exp\left(\sum_{j=1}^{l} N_{ij}x_j\right) \tag{4.4.1}$$

parametrized by a real matrix N_{ij} of full rank.

The equations of motion obtained from the above Hamiltonian are

$$\dot{x}_i = \partial H/\partial y_i, \quad \dot{y}_i = -\partial H/\partial x_i, \quad i = 1,...,l. \tag{4.4.2}$$

Remember that the size of N_{ij} is $(l+1, 1)$, whose transpose has a null vector of the form $(\bar{p},1)^+ = (p_1,...,p_1,1)^+$. Such a system is a natural generalization of the Toda Lattice. The linear transformation preserving $\langle x, y \rangle = \sum_{1}^{1} x_i y_i$, given by

$$(\bar{x}, \bar{y}) = \left(\bar{N}_x, (\bar{N}^+)^{-1}y\right)$$

with

$$N = \begin{pmatrix} \overline{N} \\ \text{last row} \end{pmatrix}$$

transforms H into

$$H = 1/2\langle \overline{M}\overline{y}, \overline{y}\rangle + \sum_{i=1}^{l} e^{x_i} + e^{-\langle \overline{p},\overline{x}\rangle} , \tag{4.4.3}$$

with

$$\overline{M} = \overline{N}\,\overline{N}^+ , \quad N^+(\overline{p},1)^+ = 0 , \quad \overline{p} = (p_1,...,p_1)^+ .$$

Using new set a_i, b_i defined by

$$a = (a_1,...,a_1)^+ = \left(e^{\overline{x}_1},...,e^{\overline{x}_1}\right)^+$$
$$a_{l+1} = e^{-\langle \overline{p},\overline{x}\rangle} , \quad b = \overline{y} ,$$

the Hamiltonian equations

$$\dot{x}_i = +\partial H/\partial \overline{y}_i , \quad \dot{y}_i = -\partial H/\partial \overline{x}_i , \quad i = 1,..., l$$

take the form

$$\dot{a}_i = a_i \overline{M}(b)_i \quad i = 1,..., l$$
$$\dot{b} = -a + \overline{p}a_{l+1} \tag{4.4.4}$$
$$\dot{a}_{l+1} = -a_{l+1}\langle \overline{M}(b), \overline{p}\rangle .$$

Clearly these equations are completely parameterized by the positive matrix

$$M = \begin{vmatrix} \overline{M} & -\overline{M}\overline{p} \\ -(\overline{M}\overline{p})^+ & \langle \overline{M}\overline{p},\overline{p}\rangle \end{vmatrix} = NN^+ ,$$

observing that $\left(\prod\limits_{i=1}^{l} a_i^{p_i} \right) a_{l+1}$ is a constant of motion, it is natural to impose the condition

that all $p_i \neq 0$. It is now easy to check that the nonlinear system (4.4.2) can also be

written as

$$\dot{x}_i = x_i \sum_{j=1}^{1+1} e_{ij} u_j \qquad |< i < 1+|$$

$$\dot{u}_i = \sum_{j=1}^{1+1} x_j e_{ji} \ , \qquad\qquad\qquad (4.4.5)$$

where $M = EE^+$, with $E = \left(e_{ij} \right)$, a real matrix of size $(1+1) \times (1+1)$ along with

$$a_i = x_i \qquad E^+ \begin{pmatrix} b \\ 0 \end{pmatrix} = u \ , \qquad H_1 = \langle \alpha, u \rangle = 0 \ .$$

For our following discussions, we will use the form given in equation (4.4.5). In fact, one
should note that $2M_{ij}/M_{jj} = a_{ij}$ is the cartan matrix.

We observe that if x_i has a pole, then $\dot{x}_i/x_i = \sum e_{ij} u_j$ has a simple pole and therefore
also some of the u_j. On the other hand, if u_i has a pole, then $\dot{u}_i = \sum x_j e_{ij}$ has a pole and
also some of the x_j. Let the vectors x and u have the following expansions

$$x = \frac{x^{(m)}}{t^m} + ...$$
$$\qquad\qquad\qquad x^{(m)}, u^{(k)} \neq 0 \ .$$
$$u = \frac{u^{(k)}}{t^k} + ...$$

Substituting in equations (4.4.5), we get

$$-m \frac{x_i^{(m)}}{t^{m+1}} + ... = \dot{x}_i = \left(\frac{x_i^{(m)}}{t^{m+n}} + ... \right) \left[\left(Eu^{(k)} \right)_i \right] \qquad (4.4.6)$$

and

$$-k \frac{u^{(k)}}{t^{k+1}} + ... = \dot{u} = \frac{E^+ x^{(m)}}{t^m} + ... + \frac{E^+ x^{(k+1)}}{t^{k+1}} + ... \ . \qquad (4.4.7)$$

Note that $m > k + 1$, otherwise $u^{(k)} = 0$. One can show that $k = 1$. For if $k > 1$ and $m > k + 1$. Then equation (4.4.7) implies

$$E^+ \left(x^{(m)} \right) = 0 \ ,$$

and hence

$$x^{(m)} = cp \ , \quad c \neq 0 \ .$$

For all i, $x_i^{(m)} \neq 0$. But equation (4.4.6) implies that as $k > 1$, $\left(E u^{(k)} \right)_i = 0$ for all i, hence $E u^{(k)} = 0$.   So $u^{(k)} = c'\alpha$, $c' \neq 0$.   Finally, equation (4.4.7) implies $E^+ x^{(k+1)} = -k u^{(k)} = -Kc'\phi$, but for any $z \langle E^+ z, \alpha \rangle = \langle z, E\alpha \rangle = 0$. We conclude that $-kc \langle \alpha, \alpha \rangle = 0$, which contradicts the hypothesis we started with. So if $k > 1$, we must have $m = k + 1$ and by (4.4.7) $-k u^{(k)} = E^+ x^{(m)}$. But because $k > 1$, equation (4.4.6) implies

$$0 = \sum_i x_i^{(m)} \left(E \overline{u}^{(k)} \right)_i = \left\langle x^{(m)}, E \overline{u}^{(k)} \right\rangle = \left\langle E^+ x^{(m)}, \overline{u}^{(k)} \right\rangle$$

$$= -k \left\langle u^{(k)}, \overline{u}^{(k)} \right\rangle = 0 \ ,$$

which is also a contradiction. So we have $k = 1$, and $m = 2$. For two column vector z and $y \, \varepsilon \, c^{l+1}$, define

$$z \cdot y = \left(z_1 y_1, z_2 y_2, \dots, z_{l+1} y_{l+1} \right)^+ \ .$$

With this notation, equation (4.4.5) can be written as

$$\left. \begin{array}{l} \dot{x} = x \cdot Eu \\ \dot{u} = E^+ x \end{array} \right\} \ . \tag{4.4.8}$$

Now set

$$\begin{array}{l} x = t^{-2} \left(x^0 + x^1 t + \dots \right) \\ u = t^{-1} \left(u^0 + ut + \dots \right) \ , \end{array} \tag{4.4.9}$$

where

$$x^0 = \left(x_i^0 \right) \neq 0 \ , \quad u^0 = \left(u_i^0 \right) \neq 0 \ .$$

One then obtains

$$-2x^0t^{-3}+...+(k+1)\chi^{k+3}t^k+...$$
$$=\left(x^0t^{-2}+...+x^{m+2}t^m+...\right)\left(Eu^0t^{-1}+...+Eu^{n+1}t^n+...\right)$$

along with

$$-u^0t^{-2}+...+(k+1)u^{k+2}t^k+...$$
$$=E^+x^0t^{-2}+...+E^+x^{k+2}t^k+... \quad . \tag{4.4.10}$$

Equating coefficients of equal powers of t, we get

$$x^0\left(Eu^0+2\right)=0$$

$$E^+x^0+u^0=0 \tag{4.4.11}$$

$$x^0\left(EE^+x^0-2\delta\right)=0$$

and

$$x\left(Eu^0+\delta\right)+x^0Eu^1=0 \tag{4.4.12}$$

$$E^+x^1=0 \quad .$$

In general, for $k\geq0$

$$x^k\left(Eu^0-(k-2)\delta\right)+x^0Eu^k=\Gamma_{k-1}=-\sum_{j=1}^{k-1}x^jEu^{k-j}$$

$$(k-1)u^k=E^+x^k$$

Also, for $k\geq2$, we find

$$x^0EE^+x^k+(k-1)x^k\cdot Eu^0-(k-1)(k-2)x^k=(k-1)\Gamma_{k-1}$$
$$(k-1)u^k=E^+x^k \quad . \tag{4.4.13}$$

If for all i, $x_i^0\neq0$, then by (4.4.11) we get

$$Eu^0=-2\delta$$

and so

$$-2\sum p_i = -2\langle \delta, p\rangle = \langle Eu^0, p\rangle = \langle u^0, E^+ p\rangle = 0 \ ,$$

which is a contradiction. Let $x_1^0, \ldots, x_s^0 \neq 0$ and $x_{s+1}^0, \ldots, x_{l+1}^0 = 0$, $1 \leq s \leq l$. For any vector z, let $\bar{z} = (z_1, \ldots, z_s)$ and $\underline{z} = (z_{s+1}, \ldots, z_{l+1})$, and for some matrix c, let $\bar{c}$ denote the upper left $s \times s$ minor. Then equations (4.4.11), (4.4.12), and (4.4.13) become

(a) $\left(\overline{Eu^0}\right) + 2\bar{\delta} = 0$

(b) $E^+ x^0 + u^0 = 0$

(c) $\left(\overline{EE^+}\right) x_0 = 2\bar{\delta}, \quad \underline{x}^0 = 0$ (4.4.14)

(a$_1$) $-\bar{x}^1 + \bar{x}^0 \cdot \left(\overline{Eu^1}\right) = 0$

(b$_1$) $x' \left[\left(\underline{Eu^0}\right) + \underline{\delta} \right] = 0$

(c$_1$) $E^+ x^1 = 0$

(a$_2$) $\bar{x}^0 \cdot \left(\overline{EE^+ x^k}\right) - k(k-1)\bar{x}^k = (k-1)\Gamma_{k-1}, \quad k > 2$

(b$_2$) $\underline{x}^k \cdot \left(\underline{Eu_0}\right) - (k-2)\underline{x}^k = \Gamma_{k-1} = -\sum_{j=1}^{k-1} x^j \cdot \left(\underline{Eu^{k-j}}\right)$

(c$_2$) $(k-1)u^k = E^+ x^k$.

Because any proper subset of the $e_i s$ is independent, $(EE^+) = \{\langle e_i, e_j\rangle\}$ is an invertible matrix and x^0, u^0 are uniquely defined. Two cases arise: $c = 0$ and $c \neq 0$. We discuss the case $c = 0$. Because $c = 0$, we have $x = 0$. Equation (a$_1$) of (4.4.14) implies $\left(Eu^1\right) = 0$, yielding $1 + 1 - s$ degrees of freedom at $k = 1$. For $k > 2$, we solve the diagonal system (b$_2$) of equation (4.4.14), producing at most $1 + 1 - s$ degrees of freedom for all $k > 2$. After putting $\underline{x}^k$ into (a$_2$), we need to solve for $\bar{x}^k$. As before,

$$\left[\bar{x}^0 \cdot (\overline{EE^+}) - k(k-1) \right] \bar{x}^k = (k-1)\overline{\Gamma}'_{k-1}, \quad k \geq 2$$

produces at most s degrees of freedom. In total, at most $2l+2-s<2l+1$ degrees of freedom may arise in the power series. Unless $s=1$, the number of degrees of freedom $<2l+1$. Incidentally, because $\overline{x}^0(\overline{EE}^+)\overline{x}^0 = \overline{x}^0(2\overline{\delta}) = 2\overline{x}_0$ holds, $k(k-1)=2$ is an eigenvalue of $x^0(\overline{EE}^+)$, picking up degrees of freedom at $k=2$ stage. So, in the case $s=1$, it may be possible to find exactly $2l+1$ degrees of freedom. Equations (b)'s and (c)'s then lead to

$$x_1^0\|e_1\|^2 = 2$$

$$x_j^0 = 0, \quad 2 \le j \le l+1 \tag{4.4.15}$$

$$u^0 = -x_1^0 e_1 = -2\|e_1\|^{-2} e_1$$

and

$$x_j^1 = 0, \quad 1 \le j \le l+1$$

$$\langle e_1, u^1 \rangle = 0 .$$

Hence u^1 is determined up to 1 degrees of freedom. An additional one comes from the fact that

$$x_1^0\|e_1\|^2 - k(k-1)$$

vanishes for $k=2$. One can then have the required degrees of freedom if

$$(Eu_0)_j = \langle e_j u^0 \rangle = k-2$$

for some $k \ge 2$. This implies that

$$a_{j1} = \frac{2\langle e_j e_1 \rangle}{\|e_1\|^2} \varepsilon - \mathbb{Z}^+$$

for $2 < j < l+1$. One can now observe that equation (4.4.15) implies that when $x_1 \to \infty$, others remain finite. The same logic can be repeated for each $x_i \to \infty$, hence one can have the general condition that

$$a_{ij} = \frac{2\langle e_i, e_j \rangle}{\|e_j\|^2} \varepsilon - \mathbb{Z}^+ .$$

This simply means that a_{ij} is the cartan matrix, and the usual compatibility conditions are automatically satisfied. By starting with a system of particles with interaction of the exponential type, it is possible to prove that for integrability the coefficient matrix, a_{ij} must be a cartan matrix associated with Lie algebra of A_1, B_1, C_1, D_1 type.

To understand the linearizability of the dynamics on a tori, consider the Euler rigid body motion. It is governed by the following equations:

$$\dot{x} = x\Lambda \cdot \Lambda x$$
$$x \in R^3 \tag{4.4.16}$$
$$\Lambda x = (\lambda_1 x_1, \lambda_2 x_2, \lambda_3 x_3) \in R^3 ,$$

where x_i are the angular momenta about the principal axes of inertia through the center of mass, and λ_i^{-1} are principal moments of inertia. The system has two conserved quantities — the total angular momentum and total energy

$$Q_1 = x_1^2 + x_2^2 + x_3^2 = C_1 \tag{4.4.17a}$$

$$Q_2 = 1/2 \left(\lambda_1 x_1^2 + \lambda_2 x_2^2 + \lambda_3 x_3^2 \right) = C_2 . \tag{4.4.17b}$$

The system evolves on the intersection of an ellipsoid and a sphere. Observe that the first equation of (4.4.16) yields

$$\frac{dx_1}{x_2 x_3} = (\lambda_3 - \lambda_2) dt . \tag{4.4.18}$$

Now solve equations (4.4.17a) and (4.4.17b) for x_2^2 and x_3^2 linearly. That is

$$x_3^2 = \frac{1}{(\lambda_3 - \lambda_2)} \left[2c_2 - \lambda_2 c_1 + (\lambda_2 - \lambda_1) x_1^2 \right]$$

$$x_2^2 = \frac{1}{(\lambda_2 - \lambda_3)} \left[2c_2 - \lambda_3 c_1 + (\lambda_3 - \lambda_1) x_1^2 \right] ,$$

and substitute in (4.4.18). This leads to an elliptic integral

$$\int_{x_1(0)}^{x_1(t)} \frac{dx_1}{\sqrt{\left(x_1^2 + \beta\right)\left(x_1^2 + \gamma\right)}} = \alpha t$$

with respect to the elliptic curve

$$y^2 = \left(z^2 + \beta\right)\left(z^2 + \gamma\right)$$

α, β, γ all depend on λ and c.

The solutions to the nonlinear system substituted in equation (4.4.16) are linear functionals of time, so $x_i(t)$ can always be expressed as theta-functional by Abel's Map. Moreoever, the circles defined by the intersection $\{Q_1 = C_1\} \cap \{Q_2 = C_2\}$ form the real part of the dimensional torus. Also, we have already seen in the Chapter 1 that a solution expressed in terms of elliptic function has an expansion in Laurent series. Indeed, the variables $x_i(t)$ in this particular case are all expressed

$$x_i(t) = t^{-1}\left(x_1^0 + x_1^1 t + x_1^2 t^2 + ...\right) .$$

One may also observe that the variables x_0/x_α, x_1/x_α, x_3/x_α, x_3/x_α, with $x_0 = 1$. $\alpha = 1, 2, 3$, also form a closed system of nonlinear equation. For example,

$$\frac{d}{dt}(x_2/x_1) = (2C_2 - \lambda_3 C_1)\frac{1}{x_1}\frac{x_3}{x_1}$$

is indeed quadratic in $1/x_1$, x_3/x_1.

The same type of linearization holds for the case of exponential interaction. The linearization is determined by the divisors on the Abelian variety. The divisors can be computed as follows. The above system of differential equations have in addition to the following trivial invariants

$$H_0 = \prod^{\ell+1} x_1 p_i \tag{4.4.19}$$

$$H_1 = \sum_1^{\ell+1} \alpha_i u_i$$

and the energy integral

$$H_2 = \frac{1}{2}\sum_1^{\ell+1}\left(u_1^2 - 2x_i\right) . \tag{4.4.20}$$

$(1-1)$ other invariants H_i, $2 \leq i \leq 1+1$. Along each divisor θ_i, each invariant H_i is finite and can be expressed after substitution by the expansions u_i and x_i as a polynomial function in β_k, $1 \leq k \leq 1+1$ and $\pi\gamma_k$, $1 \leq k \leq 1+1$. This recipe provides $\ell+2$ equations between these $2\ell+1$ parameters, defining an $\ell-1$ dimensional variety.

For example, in the case of Kac-Moody Lie algebra $a_2^{(1)}$, we have

$$H_0 = x_1 x_2 x_3 = \beta_2 \beta_3 \tag{4.4.21a}$$

$$H_1 = u_1 + u_2 + u_3 - 2\gamma_1 + \gamma_3 \tag{4.4.21b}$$

$$H_2 = 1/2\left(u_1^2 + u_2^2 - 2x_1\right) + 1/2u_3^2 - x_2 - x_3 = \left(\gamma_1^2 - 3\beta_1\right) + \gamma_3^2/2 \tag{4.4.21c}$$

$$H_3 = u_3\left(u_1 u_2 + x_1\right) + u_1 x_2 + u_2 x_3 = \left(3\beta_1 + \gamma_1^2\right)\gamma_3 + \beta_2 - \beta_3 \tag{4.4.21d}$$

where we have used the following expansions

$$x_i = \frac{2}{\|e_i\|^2}t^{-2} + \beta_i + ...$$

$$x_j = \beta_j t^{-a_{j1}} + ... j \neq i$$

$\beta \varepsilon\, C^{l+1}$ arbitrary and $x = 0$. Also,

$$u = -\frac{2}{\|e_i\|^2}e_i t^{-1} + \gamma + u^2 t + ... ,$$

where
$$\langle \gamma, e_i \rangle = 0 ,$$

whence
$$u^{(2)} = E^+ x^{(2)} = e_i \beta_i + \sum_{1 < j < l+1} e_j \beta_j .$$

These are a summarization of analysis prior to equation (4.4.22). If we eliminate $\gamma_2, \gamma_3, \beta_1, \beta_3$ between equations (4.4.21a), (4.4.21b), (4.4.21c), and (4.4.21d), we get

$$\beta_2 - \frac{H_0}{\beta^2} - p_3(\gamma_1) = 0 , \tag{4.4.22}$$

with p_3 being a cubic polynomial with coefficients depending on $H_1 (0 \leq 1 \leq 3)$. This is nothing but a hyperelliptic curve. The next important point is that the Laurent expansions of u and x provide the regularizing coordinates of the flow near the divisor θ_1 where the hyperelliptic curve is given by (4.4.22). The variables γ and β parametrize the point of intersection of the trajectory while θ_1 and x_2 measures time.

We do not go into further details, but the reader may refer to the original and pioneering article of Adler and Van Moerbeke.[89,90]

Basic conditions

We can now summarize the basic conditions of algebraic integrability. Consider the Hamiltonian system

$$x = f(x) = J \frac{\partial H}{\partial x} , \quad \alpha \, \varepsilon \, R^n , \tag{4.4.23}$$

where $J = J(x)$ is a skew symmetric matrix, so that the Poisson bracket

$$\{A, B\} = \left\langle J \frac{\partial A}{\partial x}, \frac{\partial B}{\partial x} \right\rangle \tag{4.4.24}$$

satisfies Jacobi identity. Let $H_1, \ldots H_k$ be the Casimir functions. Their gradient $\partial H_i / \partial x$ are null-vectors of J. The symplectic leaves

$$\bigcap_1^k \left\{ H_i = C_i, \ x \, \varepsilon \, R^n \right\} ,$$

which must be even dimensional. Let $n - k = 2m$. The system is called algebraically completely integrable, when:

1. The system possesses m polynomial invariants $H_{k+1}, \ldots, H_{k+m}$ in involution, i.e., $\{H_i, H_j\} = 0$ such that the invariant manifolds

$$\bigcap_1^{k+m} \{H_i = C_i, \ x \varepsilon R^n\}$$

are compact and connected. The solution of the system is then linear on the m dimensional tori

$$T_R^m = \frac{R^m}{\text{lattice}} .$$

2. The invariant manifolds, thought of as affine varieties in C^n, can be completed into complex algebraic tori, that is

$$\bigcap_1^{k+m} \{H_i = C_i, \ x \varepsilon C^n\} = T_c^m / \{\text{one or several codimensional 1 subvarieties}\} .$$

Algebraic means that T_c^m can be defined as an intersection

$$\bigcap_1^{M} \{F_i(x_0, \ldots, x_N)\} = 0 \qquad (4.4.25)$$

F_i are large numbers of homogeneous polynomials. Condition 2. means that there is an algebraic map

$$\{x_1(t), \ldots, x_n(t)\} \rightarrow \{\mu_1(t), \ldots, \mu_m(t)\} ,$$

making the following sum linear in t

$$\sum_{i=1}^{m} \int_{u_i(0)}^{u_i(t)} \omega_k = a_k t , \qquad k = 1, \ldots, m \qquad (4.4.26)$$

$\omega_1, \ldots, \omega_m$ are holomorphic differentials on some algebraic curves.

From the previous analysis, we have seen that the algebraic integrability depends crucially on the existence of integrals of motion, which in turn helps to linearize the system on the Jacobi variety. The possibility of the existence of extra integrals of motion can be ascertained by Ziglin's theorem.[54] Ziglin's theorem has been used to prove the nonintegrability of a dynamical system. The following states the theorem and then demonstrates it in some cases of interesting dynamical systems.

Ziglin's theorem

We let (Σ, ω) denote a complex analytic symplectic manifold with a Poisson bracket $\{\ ,\ \}$ for a holomorphic Hamiltonian $H: \Sigma \to C$, we let X_H be the associated vector field. The image of the maximally continued nonequilibrium integral curve $\phi(t)$ with energy $h \varepsilon C$ of X_H is a Riemann surface $\Gamma \subset \Sigma_h = H^{-1}\{h\}$. The linearized equations along $\phi(t)$ induces a linear differential equation on the reduced normal bundle $N = \{T(\Sigma_h)/\Gamma\}/T(\Gamma)$ of Γ called the reduced normal variational equation (NVE). Continuing a fixed fundamental system of solutions of the NVE around inverses of loops based at $x_0 \varepsilon \Gamma$ gives the monodromy group of the NVE as the image M of the representation $\pi_1(\Gamma, x_0) \to Sl(2, C)$. $A \varepsilon M$ is called nonresonant if no eigenvalue is a root of unity. With the stage now set, Ziglin's theorem can be stated as follows:

Assume there is a meromorphic function Γ defined in some neighborhood $V \subset \Sigma$ of Γ, functionally independent of H and satisfying $\{F, H\} = 0$ on U. Assume the monodromy group M contains a nonresonant element A. For any $B \subset M$, the commutator $[A, B] = B^{-1}A^{-1}BA$ satisfies $[A, B] = 1$ or $[A, B] = A^2$. The former one is possible if B does not admit $\neq i$ as eigenvalues. In particular, X_H has no meromorphic integral independent of H if there is a $B \varepsilon M$ such that $I \neq [A, B] \neq A^2$.

Here we give an idea of the proof based on a presentation of Yoshida.[52] Yoshida[53] considered a Hamiltonian system with n degrees of freedom of the form

$$H = \frac{1}{2}p^2 + V(q) \ .$$

$V(q)$ is assumed to be a homogeneous function of integer degree $K \neq 0, \pm 2$. Because the equations of motion are scale invariant, we can compute the Kowalevski's exponents (KEs), $\rho_1, ..., \rho_{2n}$. These exponents characterize the branching of the solution in the complex t-plane

$$q(t) = t^{-6}\left[d + \text{a Taylor series in } \left(I_1 t^{P_1}, I_2 t^{P_2} \ldots\right)\right].$$

Here $g = 2/(k-2)d = $ constant vector, I_j are integration constants. The KEs satisfy $P_1 + P_{i+n} = 2g + 1$ $(i = 1,2,\ldots,n)$. The analysis of Yoshida helps to connect Ziglin's theorem with KE, which was not in the original formulation of Ziglin. In Yoshida's analysis, an additional result can be stated as follows:

If the n numbers $(\Delta \rho_1, \Delta \rho_2, \ldots, \Delta \rho_n)$ are rationally independent, then the Hamiltonian system has no additional global analytic integral $\phi(p,q) = $ constant besides the Hamiltonian itself. We first give a recipe for the calculation of $\Delta \rho_i$.

1. Obtain a solution $C = (c_1, c_2, \ldots, c_n)$ of the algebraic equation,

$$\text{grad } V(c) = c . \tag{4.4.27}$$

   Note that $d = [-g(g+1)]^{g/2} C.$

2. Let $D^2 V(C)$ be the Hessian matrix of $V(q)$ evaluated at $q = c$. Compute the eigenvalues $(\lambda_1, \lambda_2, \ldots, \lambda_n)$ of $D^2 V(C)$.

3. The pair of $KE(\Delta \rho_i, \Delta \rho_{i+n})$ are then the root of the equation

$$\rho^2 - (2g+1)\rho + g(g+1)(1 - \lambda_i) = 0 .$$

Observe that there is a straight line solution,

$$q = c\phi(t)$$

with c given by equation (4.4.27), and ϕ satisfying

$$\ddot{\phi} + \phi^{k-1} = 0$$

or

$$\frac{1}{2}\dot{\phi}^2 + \frac{1}{4}\phi^k = \text{const} = 1/k . \tag{4.4.28}$$

The linear variational equation around this straight line solution becomes

$$\ddot{\xi}_i = -\phi(t)^{k-2} D^2 V(C)\xi_i . \tag{4.4.29}$$

After a proper normalization, this becomes

$$\ddot{\xi}_i + \lambda_i \phi(t)^{k-2} \xi_i = 0 \quad (i = 1, 2, ..., n) . \tag{4.4.30}$$

The monodromy matrix for the NVE has the form

$$M = \text{diag}\left[M(\lambda_1), M(\lambda_2), ..., M(\lambda_{m-1}) \right] ,$$

where each $M(\lambda_i)$ is 2×2 matrix and $\det M(\lambda_i) = 1$. Such a monodromy matrix is called nonresonant when its eigenvalues $(\sigma_1, 1/\sigma_1, \sigma_2, 1/\sigma_2, ...)$ have no relation of the form

$$\sigma_1^{m_1} \sigma_2^{m_2}, ..., \sigma_{n-1}^{m_{n-1}} = 1$$

for some $m_1, m_2, ..., m_{b-1}$, which are integers. Let M_1 and M_2 be two such monodromy matrices around two different loops C_1, C_2 on the Riemann surface,

$$\zeta^2 = \frac{2}{k}\left(1 - \omega^k\right) .$$

If we set $z = \{\phi(t)\}^k$, equation (4.4.30) is transformed into the hypergeometric equation,

$$Z(1-z)\frac{d^2\xi_i}{dt^2} + [C - (a+b+1)z]\frac{d\xi_i}{dt} - ab\xi_i = 0 , \tag{4.4.31}$$

with
$$a + b = 1/2 - 1/k$$
$$ab = -\lambda_i/2k$$
$$c = 1 - 1/k .$$

Using the basic expression for the monodromy matrix of equation (4.4.31), $M(\gamma_0) = M(\gamma_0, k, \lambda_i), \ M(\gamma_1) = M(\gamma_1, k, \lambda_i)$, one gets

$$M_1(\lambda_i) = \left\{ M(\gamma_0) M(\gamma_1) \right\}^{4k}$$
$$M_2(\lambda_i) = \left\{ M(\gamma_1) M(\gamma_0) \right\}^{4k} ,$$

where

$$M(\gamma_0) = \begin{pmatrix} 1 & B-C \\ 0 & C \end{pmatrix}$$

$$M(\gamma_1) = \begin{pmatrix} AB/C & 0 \\ 1-A/C & 1 \end{pmatrix},$$

with

$$A = e^{-2\pi i a}, \quad B = e^{-2\pi i b}, \quad c = e^{-2\pi i c}.$$

Observe that the product $M(\gamma_0)M(\gamma_1)$ is diagonalized unless $A = B$ by

$$u = \begin{pmatrix} 1 & B-C \\ -1 & C-A \end{pmatrix}, \tag{4.4.32}$$

as

$$u^{-1}M(\gamma_0)M(\gamma_1)u = \begin{pmatrix} A & 0 \\ 0 & B \end{pmatrix},$$

so that

$$u^{-1}\{M(\gamma_0)M(\gamma_1)\}^{4k}u = \begin{pmatrix} A^{4h} & 0 \\ 0 & B^{4k} \end{pmatrix}.$$

It is interesting to note that

$$u^{-1}[M(\gamma_1)M(\gamma_0)]^{4k}u$$

$$= u^{-1}M(\gamma_0)^{-1}u \cdot u^{-1}[M(\gamma_0)M(\gamma_1)]^{4k}u \cdot u^{-1}M(\gamma_0)u.$$

Yoshida showed that if both M_1 and M_2 are nonresonant and if none of the pairs $M_1(\gamma_1)$ and $M_2(\gamma_1)$ commute, then the system (4.4.28) has no additional analytic integral besides the Hamiltonian. In the above expressions for the monodromy matrices M, (γ_0, γ_1) stand for the two loops C_1 and C_2 chosen freely on the Riemann surface given above. It is always possible to choose another set C_1 and C_2 so that C_1 and C_2 are mapped to

$$C_1 \longrightarrow (\gamma_0 \gamma_1)^{4k}, \quad C_2 \longrightarrow (\gamma_1 \gamma_0)^{4k}.$$

The matrices $M_1(\lambda_1)$ and $M_2(\lambda_i)$ have a common trace equal to $2\cos(2\pi\,\theta i)$ such that

$$\theta_i = (k-2)^2 + 8k_i \ .$$

From the analysis given in equation (4.4.32), it is clear that the matrices $\{M(\gamma_0)M(\gamma_1)\}^{4k}$ and $\{M(\gamma_1)M(\gamma_0)\}^{4k}$ will commute if they are diagonalized simultaneously. By requiring that the off-diagonal elements be 0, one obtains the condition that traces $M_{1,2}(\lambda_i) = \pm2$ or $K = \pm2$. Because the eigenvalue of $M_{1,2}(\lambda_i)$ is $\sigma_j = e^{2\pi}\theta_j$ and $\theta_n = 3k-2$ an integer, M_1 and M_2 are nonresonant whenever the n numbers θ_i $(i = 1, 2,...,n)$ are rationally independent. One should also note that none of the pair $M_1(\lambda_i)$ and $M_2(\lambda_i)$ can commute unless $k = \pm2$.

Now consider some aspects of Kowalevski's exponents for the above system. Consider an autonomous system

$$x_i = F_i(x_1, x_2,..., x_N)$$

invariant under a scaling

$$t \to \alpha^{-1}t, \quad x_i \to \alpha^{g_i}x_i \quad (i = 1,...,N) \ .$$

Fix a solution $d = (d_1, d_2,..., d_N)$ of the algebraic equations

$$F_i(d_1, d_2,..., d_N) = -g_i d_i \ .$$

We have already discussed these in Chapter 3. The eigenvalues of $K = K_{ij}$,

$$K_{ij} = \left(\frac{\partial F_i}{\partial x_j}\right)_{x-d} + \delta_{ij}g_i \ ,$$

are known as KE. Note that the characteristic polynomial $K(\rho) = \det|I\rho - K|$ used to compute these is

$$K(\rho) = \det\left|(\rho - g)(\rho - g - 1)I + D^2V(d)\right| \ ,$$

where d is a solution of

$$\text{grad } V(d) = -g(g+1)d \ .$$

The solution of this equation and that of grad $V(C) = C$ are related by $d = [-g(g+1)]^{g/2}C$, and because the matrix $D^2V(q)$ is a homogeneous function of degree $K-2$, we get

$$D^2V(d) = -g(g+1)D^2V(C) \ .$$

The roots of $K(\rho) = 0$ is given by the quadratic equation

$$\rho(2g+1) + g(g+1)(1-\lambda_i) = 0 \ .$$

The whole set up can be more clearly appreciated if we study the example of the three-dimensional Yang-Mills potential considered by Steeb *et al.* Here,

$$V(q) = q_1^2 q_2^2 + q_2^2 q_3^2 + q_3^2 q_1^2 \ .$$

Now the solution C is

$$C = \left(2^{-1/2}, 2^{-1/2}, 0\right) \ .$$

$(\lambda_1, \lambda_2, \lambda_3) = (2, -1, 3)$ and consequently $(\Delta_1, \Delta_2, \Delta_3) = (\ 17, \ -7, \ 5)$. Because these values are rationally independent, the three-dimensional Yang-Mills system has no more global analytic integral.

The above discussion makes it very clear that the study of the monodromy data is very important in determining the algebraic integrability of the nonlinear system. But one main difficulty associated with it is that this mode of analysis is not easily extended to the case of nonlinear partial differential systems. In such a case, one can take recourse to the ARS reduction and then analyze the dynamical system under consideration.

4.5 Painleve analysis of some special systems

Previous discussions have detailed the basics of Painleve analysis and how it can be applied. Of course, there are some problems that may require special attention, and such cases always differ from the usual prescription. One such case is given by the coupled KdV problem that was studied by Gurses and Nutku. They obtained the inverse scattering equations and the corresponding soliton solutions. But Roy Chowdhury and Chanda[98] considered the Painleve test for such equations. The equation under consideration can be written as,

$$u_t + 6uu_x + u_{xxx} - \lambda_x u_{xx} = 0$$

$$\lambda_t + 2u\lambda_x + 2ku_x = 0 .$$

(4.5.1)

The Painleve analysis starts with the usual assumption:

$$u = \sum_j u_j \phi^{j+\alpha}$$

$$\lambda = \sum_j \lambda_j \phi^{j+\beta}$$

(4.5.2)

$\phi = 0$ being the singular manifold.

Here the leading order analysis makes the difference. The leading power of λ turns out to be 0, and that is why we set

$$u = u_0 \phi^{-2}$$

$$\lambda = n_0 \log \phi + \lambda_0 .$$

(4.5.3)

This is one branch. The other possibility is given as

$$u = u_0 \phi^{\frac{2}{1+k}}$$

$$\lambda = n_0 \log \phi + \lambda_0 ,$$

(4.5.4)

where in the first case

$$u_0 = -(2+k) , \quad n_0 = 2K$$

$$\lambda_0 = \text{undetermined}$$

and in the second one

$$\alpha = \frac{2}{k+1} \qquad n_0 = -K\alpha .$$

But because we demand that $\alpha < 0$, it imposes some constraint on K. Here, u_0 remains arbitrary. For the study of resonances, we set

$$\lambda = n_0(t)\ln\phi + \sum_{j=0}^{\infty}\lambda_j(t)\phi^j$$

$$u = \sum u_j(t)\phi^{-2+j} \qquad\qquad (4.5.5a)$$

$$\phi = x + \psi(t) ,$$

whence from equation (4.5.1) we arrive at

$$\sum_{j=0}^{\infty}u_{jt}\phi^{j-2} + \sum_{j=0}^{\infty}u_j(j-2)\phi^{j-3}\psi_t + 6\sum_j\sum_k u_ju_k(k-2)\phi^{j+k-5}$$

$$+\sum u_j(j-2)(j-3)(j-4)\phi^{j-5} - n_0\sum u_j(j-2)(j-3)\phi^{j-5}$$

$$-\sum_{j=0}^{\infty}\sum_{k=0}^{\infty}k(j-2)(j-3)\lambda_ku_j\phi^{j+k-5} = 0$$

and

$$n_0\phi_t\phi^{-1} + \sum\lambda_{jt}\phi^j + \sum\lambda_j j\phi^{j-1} + 2n_0\sum u_j\phi^{j-3}$$

$$+\sum_j\sum_k u_j\lambda_k k\phi^{j+k-3} + 2k\sum u_j(j-2)\phi^{j-3} = 0 .$$

From this we get the condition that $\det\Delta = 0$, where Δ is given by

$$\Delta = \begin{pmatrix} \gamma^3 - (2k+q)\gamma^2 + (4k+14)\gamma + 12k + 24 & 6(2+k)\gamma \\ k\gamma & -(2+2k)\gamma \end{pmatrix} \qquad (4.5.5b)$$

for the nonzero values of λ_j, u_j. This equation leads to the resonance positions that are given by

$$\gamma(\gamma+1)(\gamma-6)[\gamma-(2k+4)]=0 , \qquad (4.5.6)$$

that is at $\gamma = 0, -1, 6, 2k+4.$

It may be noted that the resonance position $2k+4$ will be of interest if $2k+4$ is an integer. To check for the arbitrariness of the coefficients, we consider the various coefficients of ϕ^{-j}.

For $j=1$, we get

$$(14k+30)u_1 + 6(2+k)\lambda_1 = 0$$

$$ku_1 - (2+k)\lambda_1 = 0 .$$

These equations are consistent only when

$\alpha)$ $k = -3/2$

$\beta)$ $u_1 = \lambda_1 = 0 ,$ all k.

Because we do not want to fix the value of k, we proceed with the second alternative. Picking up coefficients of ϕ^{-3} and ϕ^{-1}, we get

$$-2u_0\psi_t - 12u_0u_2 - 6u_1^2 - 12\lambda_2u_0 - 2\lambda_1u_1 = 0$$

$$n_0\psi_t + 2n_0u_2 + 4u_0\lambda_2 + 2u_1\lambda_1 = 0 .$$

These equations together with $u_1 = \lambda_1 = 0$ lead to

$$u_2 = -\frac{1}{6}\left(\frac{2k+1}{k+1}\right)\psi_t \qquad (4.5.7)$$

$$\lambda_2 = \frac{k}{6(k+1)}\psi_t .$$

Proceeding similarly with the coefficients of ϕ^{-2} and ϕ^0, we get for $j=3$

$$u_3 = -\frac{1}{4(2k+1)}\lambda_{0t}$$

$$\lambda_3 = \frac{1}{12(2k+1)}\lambda_{0t} . \qquad (4.5.8)$$

Not that these become undefined for $k = -1/2$.

On the other hand, for $j = 4$ we get

$$u_4 = -(3/10k)[\psi_t + 2u_2]\lambda_2$$
$$\lambda_4 = (\lambda_2/20(2+k))[\psi_t + 2u_2] .$$

And for $j = 5$,

$$u_5 = (g + 3h)/(12k - 6)$$
$$\lambda_5 = (101g + 48kh + 6k)/10(2+k)(12k-6) ,$$

where
$$-g = +u_{2t} + u_3\psi_t + 6u_2u_3$$
$$-h = \lambda_{2t} + 3\lambda_3\psi_t + 6u_2\lambda_3 + 6u_3\lambda_2 .$$

One can observe that the consistency condition is satisfied if

$$\lambda_0 = -1/3\ln\psi_t + \text{constant} .$$

Lastly, for $j = 6$ we get two more equations:

$$-36k\,u_6 + 36(k+2)\lambda_6 = -u_{3t} - 2u_4\psi_t - 6u_3^2 - 12u_2u_4 + 4\lambda_2u_4$$
$$12k\,u_6 - 12(k+2)\lambda_6 = -\lambda_{3t} - 4\lambda_4\psi_t - 4u_4\lambda_2 .$$

For consistency we should have the equality

$$(u_3 + 3\lambda_3)_t + 6u_3(u_3 + 3\lambda_3) + u_4(2\psi_t + 12u_2 + 8\lambda_2) + \lambda_4(12\psi_t + 24u_2) = 0 .$$

Using the previously obtained expressions for $u_4, u_2, u_3, \lambda_2, \lambda_3, \lambda_4$, one can check that this equation is identically satisfied. Therefore, we are left with only one equation:

$$12(k+2)u_6 = 12(k+2)u_6 + \lambda_{3t} + 4\lambda_4\psi_t + 4u_4\lambda_2 + 6u_3^2 + 8u_2\lambda_4 ,$$

so u_6 is arbitrary at $j = \sigma$.

Here we have a situation where despite the fact that the expansion for one of the nonlinear variable starts with a logarithm, we have the requisite number of resonances and the consistency conditions can be satisfied except for some special values of K. Remember also that there exists a second branch given by equation (4.5.4).

For the second branch, we set

$$u = \sum_{j=0}^{\infty} u_j(t)\phi^{\frac{2}{k+1}} + j$$

$$\lambda = -\frac{2k}{k+1}\ln\phi + \sum_{j=0}^{\infty} \lambda_j(t)\phi^j \binom{k \neq -1}{0,-2}. \tag{4.5.9}$$

Substituting in equation (4.5.1), we get

$$\sum u_{jt}\phi^{q+j} + \sum u_j(q+j)\phi^{q+j-1}\psi_t$$

$$+6\sum u_j u_m(q+m)\phi^{(q+j+m-1)} + \sum u_j(t)(q+j)$$

$$\times(q+j-1)(q+j-2)\phi^{q+j-3} + 2q\sum u_j(t)(q+j)(q+j-1)\phi^{q+j-3}$$

$$-\sum \lambda_m m u_j(q+j)(q+j-1)\phi^{q+j+m-3} = 0$$

and

$$-2q\phi^{-1}\psi_t + \sum \lambda_{jt}\phi^j + \sum \lambda_j(t)j\phi^{j-1}\psi_t - 4q\sum u_j\phi^{q+j-1}$$

$$+2\sum \lambda_m u_j m\phi^{q+j+m-1} + 2k\sum u_j(q+j)\phi^{q+j-1} = 0 . \tag{4.5.10}$$

The equation determining the resonances turns out to be

$$\gamma^2(\gamma+1)\left[\gamma\left(k^2+2k+1\right)+\left(2-2k^2\right)\right]=0 ,$$

whose roots are

$$\gamma = 0, \quad 0, \quad -1, \quad 2\frac{k-1}{k+1}. \tag{4.5.11}$$

For an effective resonance position, k should have a value so that

$$2\frac{k-1}{k+1} \quad \text{is an integer}$$

or

$$k = \frac{2+n}{2-n}, \quad n > 0 \quad \text{integer.}$$

We observe that for $n = 3, 4$, k has the value -5 and -3, respectively. One can now proceed as before to test the consistency and arbitrariness of the coefficients. There is an equation that is related to the equation by a simple transformation. This was discussed by Roy Chowdhury and Paul.[99] It can be written as

$$\rho u_{xxx} - k\rho_x u_{xx} + 6\rho u_{ux} + \rho u_t = 0$$
$$\rho_t + 2\rho u_x + 2u\rho_x = 0 \ . \tag{4.5.12}$$

If we set

$$u = \sum u_j(t)\phi^{\alpha+j}$$
$$\rho = \sum {}_j(t)\phi^{\beta+j} \ ,$$

we find that

$$\alpha) \ \ \alpha = -2, \ \beta = 2$$
$$\beta) \ \ \alpha = 2/1+k, \ \beta = -2/1+k \ .$$

One of the exponents remains positive, so one may note that the simple transformation

$$\lambda = K(\ln \rho)$$

converts this set (4.5.12) into (4.5.1). For the first case (α), the resonance positions are given by

$$\gamma(\gamma + 1)(\gamma - 6)(\gamma - 4 - 2k) = 0 \ .$$

The resonance positions are identical to those of equation (4.5.1). At this point, we should add that the logarithmic term in the Painleve expansion is not always removable.

Dissipative and forced anharmonic system

To demonstrate the use of logarithmic term in the expansion, we consider the situation of a very general anharmonic oscillator that Steeb *et al.* looked at in detail.

In their paper (57), they considered four equations written as follows:

(a) $\ddot{x} + bx + cx^3 = 0$

(b) $\ddot{x} + bx + cx^3 = K$

(c) $\ddot{x} + a\dot{x} + bx + cx^3 = 0$ (4.5.13)

(d) $\ddot{x} + a\dot{x} + bx + cx^3 = f(t)$,

where $a, b, c, K \in \mathbb{R}$, $a > 0$, $c > 0$. The term f, represents an external forcing which can be expanded into a Taylor series,

$$f(t) = \sum f_j t^j .$$
                                                          (4.5.14)

A special and important case is given by $f = K\cos(\Omega t)$. Consider equation (a) in the complex domain, and set

$$\omega(t) \sim a_0 (t - t_1)^k .$$

This leads to $k = -1$, $a_0 \neq 0$. There is only one branch. The resonances are $\gamma = -1$ and $\gamma = 4$. Now insert the full series,

$$\omega(t) = \sum_{j=0}^{\infty} a_j (t - t_1)^{j-1} ,$$
                                                           (4.5.15)

so that a simple computation yields

$$a_0 = (-2/c)^{1/2} , \quad a_1 = 0 , \quad a_2 = -b/(3ca_0) .$$

And for $m \geq 3$, we obtain

$$[(m-1)(m-2) - 6]a_m + ba_{m-2} + C \sum_{j+k+n=m-1} a_j a_k a_n = 0 .$$
                                                     (4.5.16)

The solution of this recursion relation yields $a_3 = 0$.   At $\gamma = 4$, we get $0 \cdot a_4 = 0$.   The situation is consistent, and the Painleve property is satisfied. In case of equation (b), we have almost the same situation, only that $a_0 = (-2/c)^{1/2}$, $a_1 = 0$, $a_2 = -b/(3ca_0)$, $a_3 = -k/4$, and at $\gamma = 4$ we again get $0 \cdot a_4 = 0$. But now consider equation (c). This is equation (a) with an added damping term. Again the resonances are the same, and we get

$$a_0 = (-2/c)^{1/2} ,$$

$$a_1 = a/(3ca_0) ,$$

$$a_2 = \left(1/\left(6a^2\right) - b\right)\Big/(3ca_0)$$

$$a_3 = (1/(9a) - a/(2b))/(3ca_0) .$$

But at the resonance $\gamma = 4$, we get

$$0 \cdot a_4 = 4a \, a_3 .$$

This is inconsistent, so the Painleve property is lost! In such situations we can use an expansion involving logarithmic psi series. That is, we set

$$x(t) = (t - t_1)^{-1} \sum_{k=0}^{\infty} \sum_{j=0}^{\infty} a_{jk}(t - t_1)^j \left\{ (t - t_1)^4 \ln(t - t_1) \right\}^4 , \tag{4.5.17}$$

where

$$a_{00} = (-2/c)^{1/2} , \qquad a_{10} = a/(3c \, a_{00})$$

$$a_{20} = 1/(3ca_{00})\left(1/6a^2 - b\right)$$

$$a_{30} = a/(3c \, a_{00})\left(a^2/9 - b/2\right) .$$

The expansion coefficients a_{01} and a_{40} are determined by the equations

$$0 \cdot a_{01} = 0$$

$$0 \cdot a_{40} = 5a_{01} + 2a \, a_{30} + ba_{20} + 6c \, a_{00}a_{10}a_{30} + 3c \, a_{10}^2 + 3ca_{00}a_{20}^2 .$$

Consequently, if we set

$$5a_{01} = -\left(2aa_{30} + ba_{20} + 3c \, a_{00}a_{20}^2 + 6c \, a_{00}a_{10}a_{30} + 3c \, a_{10}^2a_{20}\right) ,$$

a_{40} can be chosen arbitrarily.   Therefore, we can bypass the difficulty at the $j = 4$ resonance by introducing the psi series.   We have seen that two completely different situations can have logarithmic terms in the Painleve expansion.   One comes from the leading order exponent 0 and the other from the inconsistency of the equations involving the expansion coefficients.   At this time, we point out that such logarithmic type psi series was successfully used in the Painleve analysis of the famous Lorentz system, which was almost the first dynamical system to exhibit the phenomena of chaos.   This analysis was due to Levine and Tabor.   Here we will just mention some salient features of their elegant paper to acquaint the reader with another use of psi series.

Lorentz system

The Lorenz system was initially studied as a model equation for the description of the atmospheric instability.   The phenomenon of chaos was predicted from the numerical analysis of the system.   This equation is written as

$$\dot{X} = \sigma(Y - X)$$
$$\dot{Y} = -XZ + RX - Y \qquad\qquad\qquad (4.5.18)$$
$$\dot{Z} = XY - BZ \; ,$$

where σ, R, $B \varepsilon R^+$.   The Painleve expansion starts with the leading order substitution,

$$X = a_0(t - t^*)^\alpha$$
$$Y = b_0(t - t^*)^\beta$$
$$Z = c_0(t - t^*)^\gamma \; .$$

It is easily seen that

$$\alpha = -1 , \qquad \beta = -2 , \qquad \gamma = -2$$

and

$$a_0 = 2i , \qquad b_0 = -2i / \sigma , \qquad c_0 = -2 / \sigma \; ,$$

so one starts with the ansatz

$$X = \sum_{j=0}^{\infty} a_j (t - t^*)^{j-1}$$

$$Y = \sum_{j=0}^{\infty} b_j (t - t^*)^{j-2} \qquad (4.5.19)$$

$$Z = \sum_{j=0}^{\infty} c_j (t - t^*)^{j-2} .$$

The recursion relation for the expansion coefficients turns out to be

$$\begin{pmatrix} j-1 & -\sigma & 0 \\ -2/\sigma & j-2 & 2i \\ 2i/\sigma & -2i & j-2 \end{pmatrix} \begin{pmatrix} a_j \\ b_j \\ c_j \end{pmatrix} = \begin{pmatrix} -\sigma a_{j-1} \\ R a_{j-1} - b_{j-1} - \sum a_q c_{j-q} \\ -B c_{j-1} + \sum a_q b_{j-q} \end{pmatrix} \qquad (4.5.20)$$

which is represented as

$$M \cdot V = W$$
$$V = \left(a_j, b_j, c_j \right)^t$$

and

$$W = \left(-\sigma a_{j-1}, \ R a_{j-1} - b_{j-1} - \sum a_q c_{j-q}, \ -B c_{j-1} + \sum a_q b_{j-q} \right)^t .$$

The determinant of the matrix M vanishes at $j = 2$ and 4. The compatibility conditions at $j = 2$, and 4 turns out to be

$$j = 2 : 6\sigma^2 - B\sigma - 2\sigma = B(B-1) \qquad (4.5.21)$$

$$j = 4 : \sigma^2 (21 - 3B) + B2(13 + 2\sigma) + 44\sigma + 4B - 55B\sigma - 17 = 0 \qquad (4.5.22)$$

along with the condition

$$(1 + 2B - 3\sigma)^2 \left[\tfrac{4}{3}(1 - B + 3\sigma)^2 + \tfrac{4}{3}(1 - B + 3\sigma)(5 + 9\sigma + 7B) \right.$$
$$+ 8\sigma(4 + 2B - 6\sigma) \right] + 8B(1 - B + 3\sigma)(1 + 2B - 3\sigma)(4 + 2B - 6\sigma)$$
$$= (1 - R)[72B\sigma(5 + B - 3\sigma)] . \qquad (4.5.23)$$

Solution of the equations (4.5.21), (4.5.22), and (4.5.23) reveals the following cases of Lorentz system, which are rarely completely integrable:

(i) $\sigma = 1/2$, $B = 1$, $R = 0$

two integrals of motion

solutions in terms of Jacobi functions

(ii) $\sigma = 1$, $B = 2$, $R = 1/9$

one integral of motion

It reduces to second Painleve transcendent.

(iii) $\sigma = 1/3$, $B = 1$, $R =$ arbitrary

It reduces to third Painleve transcendent.

(iv) $B = 1$, $R = 0$, $\sigma =$ arbitrary

There is one time-dependent integral of motion, but it loses the Painleve property.

(v) $B = 2\sigma$, $R =$ arbitrary

The same situation as (iv).

But if any one of conditions (4.5.21), (4.5.22), and (4.5.23) fails, then logarithmic corrections are to be introduced. We write the modified expansions as follows:

$$X = \sum_{j=0}^{\infty} \sum_{k=0}^{\infty} a_{jk} \tau^{j-1} \left(\tau^2 \ln \tau \right)^k \qquad (4.5.24)$$

$$Y = \sum_{j=0}^{\infty} \sum_{k=0}^{\infty} b_{jk} \tau^{j-2} \left(\tau^2 \ln \tau \right)^k \qquad (4.5.24b)$$

$$Z = \sum_{j=0}^{\infty} \sum_{k=0}^{\infty} c_{jk} \tau^{j-2} \left(\tau^2 \ln \tau \right)^k , \qquad (4.5.24c)$$

where $\tau = t - t^*$. This substitution will work when none of conditions (4.5.21), (4.5.22), and (4.5.23) are satisfied. The recursion relation obviously gets modified

$$\begin{pmatrix} 2k+j-1 & -\sigma & 0 \\ -2/\sigma & 2k+j-2 & 2i \\ 2i/\sigma & -2i & 2k+j-2 \end{pmatrix} \begin{pmatrix} a_{jk} \\ b_{jk} \\ c_{jk} \end{pmatrix} = s_{jk} . \qquad (4.2.25)$$

S_{jk} is a vector whose elements depend are $a_{j-2,k+1}$, $a_{j-1,k}$, $c_{j-2,k+1}$, $b_{j-q,0}$, etc. The logarithmic corrections may now be understood as follows. Had we expanded the solution about a singularity in a Laurent series without satisfying the conditions (4.5.21), (4.5.22), and (4.5.23), the coefficients (a_2, b_2, c_2) and (a_4, b_4, c_4) would be fixed at some values. But because the matrix is singular at $j = 2$ and $j = 4$, these coefficients remain arbitrary. On the other hand, the matrix in (4.5.25) is singular at six orders:

(i) $k = j = 0$

(ii) $2k + j = 2$ or
      $k = 1, \ j = 0$
      $k = 0, \ j = 2$

(iii) $2k + j = 4$ or
      $k = 2, \ j = 0$
      $k = 1, \ j = 2$
      $k = 0, \ j = 4$.

The arbitrariness at $j = 2$ is restored by fixing the coefficients at the logarithmic terms at $j = 0$, $k = 1$ so that

$$V_{20} \cdot S_{20} = 0 , \qquad\qquad (4.5.26)$$

where V_{20} is the null vector of the matrix (4.2.25). In other words, we set (a_{01}, b_{01}, c_{01}) so that

$$V_{20} \cdot S_{20} = \begin{pmatrix} -2i/\sigma \\ 0 \\ 1 \end{pmatrix} \cdot \begin{pmatrix} -a_{01} - \sigma a_{10} \\ -b_{01} - a_{10}c_{10} + Ra_{00} - b_{10} \\ -c_{01} + a_{10}b_{10} - Bc_{10} \end{pmatrix} = 0 .$$

Therefore,

$$a_{00} = 2i \qquad a_{10} = (-i/3)(1 + 2B - 3\sigma)$$
$$b_{00} = -2i/\sigma \quad b_{10} = +2i \qquad\qquad\qquad\qquad (4.5.28)$$
$$c_{00} = -2/\sigma \quad c_{10} = (2/3\sigma)(1 - B + 3\sigma) .$$

The condition determining (a_{01}, b_{01}, c_{01}) turns out to be

$$\begin{pmatrix} 1 & -\sigma & 0 \\ -2/\sigma & 0 & 2i \\ 2i/\sigma & -2i & 0 \end{pmatrix} \begin{pmatrix} a_{01} \\ b_{01} \\ c_{01} \end{pmatrix} = \begin{pmatrix} 0 \\ 0 \\ 0 \end{pmatrix} , \qquad (4.5.29)$$

so that

$$(a_{01}, b_{01}, c_{01})^t = \lambda(\sigma, 1, -1)^t .$$

λ is a free parameter. At $j = 0$, $k = 2$ a new parameter γ will be set so that there is an arbitrary coefficient at $j = 4$, $k = 0$. The form of the coefficient vector at $j = 0$, $k = 2$ is easily determined to be

$$\begin{pmatrix} a_{02} \\ b_{02} \\ c_{02} \end{pmatrix} = \begin{pmatrix} \gamma \\ 3\gamma/\sigma \\ 2i\gamma/\sigma \end{pmatrix} + \begin{pmatrix} 0 \\ 0 \\ \sigma/2 \cdot \lambda^2 \end{pmatrix} .$$

γ is fixed by the condition

$$V_{40} \cdot S_{40} = \begin{pmatrix} 0 \\ i \\ 1 \end{pmatrix} \cdot S_{40} = 0 .$$

Finally, we have

$$\lambda = (2i/9\sigma)\left[B(B-1) + 2\sigma + B\sigma - 6\sigma^2 \right]$$

$$\gamma = (\lambda\sigma/5)\left[8/9(B-1)^2 + 1/9(B-1)(17-28\sigma) + \sigma/3(2\sigma-1) \right] .$$

We have achieved a formal power series representation of the solution near a singular point, which turns out to be a logarithmic branch point the terms in the partial sums.

$$X = 1/\tau \sum_{k=0}^{\infty} a_{0k} \left(\tau^2 \ln \tau \right)^k$$

$$Y = 1/\tau^2 \sum_{k=0}^{\infty} b_{0k} \left(\tau^2 \ln \tau \right)^k$$

$$Z = 1/\tau^2 \sum_{k=2}^{\infty} c_{0k} \left(\tau^2 \ln \tau \right)^k$$

were originally used by Tabor and Weiss as a representation of the Lorenz variables. These terms will be dominant in the full series if and only if

$$\left|\tau^2 \ln \tau\right| << |\tau| ,$$

which can be satisfied by choosing a high enough Riemann sheet. From the general recursion given in equation (4.2.25), it can be demonstrated that the coefficients of these partial sums, namely (a_{0k}, b_{0k}, c_{0k}), are independent of the rest of the coefficients. They satisfy a relation that does not involve any coefficient (a_{ik}, b_{ik}, c_{ik}) with $i \neq 0$. Therefore, we may recast X, Y, Z as

$$X = \lim_{\tau \to 0} 1/\tau \, \theta_0(z)$$

$$Y = \lim_{\tau \to 0} 1/\tau^2 \, \phi_0(z)$$

$$Z = \lim_{\tau \to 0} 1/\tau^2 \, \psi_0(z) ,$$

where the partial sums are treated as Taylor series for $\theta_0(z)$, $\phi_0(z)$, $\psi_0(z)$, and the limit is taken with z fixed, where $z = \tau^2 \ln \tau$. This leads to the following coupled set of nonlinear differential equation:

$$2z\theta_0'(z) - \theta_0(z) - \sigma\phi_0(z) = 0$$

$$2z\phi_0'(z) - 2\phi_0(z) + \theta_0(z)\psi_0(z) = 0$$

$$2z\psi_0'(z) - 2\psi_0(z) - \theta_0(z)\phi_0(z) = 0 , \qquad (4.5.30)$$

which is actually referred to as the rescaled equations of motion. One may note that this rescaled system is exactly integrable regardless of the values of the parameters σ, k, and B. Equation set (4.5.30) can be directly integrated, and one can show that

$$\theta_0(z) = yf(y) , \quad f(y) = a/ Sn(ia \, y/2 \, jm)$$

$$a^2 = 12i\sigma , \quad m = \frac{3\lambda\sigma + \sqrt{2\lambda^2\sigma^2 + 20i\gamma}}{3\lambda\sigma - \sqrt{2\lambda^2\sigma^2 + 20i\gamma}} ,$$

sn being the Jacobi elliptic function and $y = z^{1/2}$. It is important to note that $f(y)$ turns out to be a solution of the following differential equation:

$$f''(y) - 3i\lambda\sigma \, f(y) + \frac{1}{2} f^3(y) = 0 .$$

The solutions $f(y)$ exhibit a regular, rectangular array of poles in the y-plane, owing to the multivaluedness of $y = \tau(\ln\tau)^{1/2}$. Every pole maps into an infinite cluster of logarithmic branch point, accumulating at the point $\tau = 0$. The mapping can be determined explicitly if we solve the equation

$$a\tau(\ln\tau)^{1/2} = ay_{\text{pole}} = lk(m) + ink'(m) .$$

y_{pole} is a position of a pole in the y-plane lattice indexed by integers (e,n) and $a^2 = 12i\lambda a$. A detailed numerical investigation of such mapping and the clustering of singularities were done by Levine and Tabor.[100] We do not go into those details but refer the reader to the original literature.

Our above discussion points to an important aspect of the Painleve test. It is the use of logarithmic terms in the expansion. There may be two types of situations. There may be finite number of terms involving logarithm, or one may require infinite series involving logarithm – the psi-series.

Modified Boussinesq equation

Now consider the Modified Boussinesq equation (MBqE) and we show that the Painleve series, including the logarithmic term, is needed even in this case. The equation is written as

$$\frac{1}{3} u_{tt} - u_t u_{xx} - \frac{3}{2} u_x^2 u_{xx} + u_{xxxx} = 0 .$$ (4.5.31)

As usual, we seek a solution in the form

$$u(x,t) = \phi^p \sum_{j=0}^{\infty} u_j(t)\phi^j ,$$

with

$$\phi(x,t) = x + \psi(t) = 0$$

as the singular manifold. The leading order analysis tells us that $p = 0$. And by equating coefficient of ϕ^{j-4}, we determine the recursion relation for the coefficients u_j. It is observed that u_0, u_1, u_2, u_3, and are all arbitrary, so we have five arbitrary functions rather than four, for a fourth order equation. This is explained by the fact that $p = 0$ represents a regular solution, for which the position of the singular manifold is irrelevant. Therefore, we may set $(t) = 0$. Here we have a situation similar to the case of equation (4.5.1), for which one of the leading exponents also turned out to be 0. So we now try to obtain a solution in the form

$$u(x,t) = v_0(t)\ln\phi + \sum_{j=0}^{\infty} u_j(t)\phi^j .$$

The leading order analysis leads to

$$V_0(t) = 2 ,$$

and equating coefficients of powers of ϕ

$$\phi^{-3} : 3u_1 + \frac{d\psi}{dt} = 0$$

$$\phi^{-2} : 12u_2 + 2\frac{du_0}{dt} = (d\psi / dt)^2$$

$$\phi^{-1}\phi^{-1} : \frac{d^2\psi}{dt^2} + 3\frac{du_1}{dt} = 0$$

$$\phi^{j-4} : \frac{1}{3}\frac{d^2}{dt^2}(u_{j-4})\frac{1}{3}(j-3)\left\{2\frac{d\psi}{dt}\frac{du_{j-3}}{dt}\right\}$$

$$+ \frac{d^2\psi}{dt^2}u_{j-3}\frac{1}{3}(j-3) + \frac{1}{3}(j-2)(j-3)u_{j-2}\left(\frac{d\psi}{dt}\right)^2$$

$$+ j(j-1)(j-2)(j-3)u_j - 2(j-1)(j-2)u_{j-1}$$

$$\times\frac{d\psi}{dt} + 2\frac{du_{j-2}}{dt} + 2(j-1)u_{j-1}\frac{d\psi}{dt}$$

$$= \frac{1}{2}(j-3)\sum_{k=0}^{j}\sum_{m=0}^{k}v_m v_{k-m}v_{j-k}$$

$$+ \sum_{k=4}^{j}(j-k+1)(k-k+2)\left\{\frac{du_{n-4}}{dt} + (k-3)\frac{d\psi}{dt}u_{k-3}\right\}u_{j-k+2}$$

for $j \geq 4$, $v_j = ju_j$ for $j \geq 1$. It follows that

$$(j+1)(j-3)j(j-4)u_j = F_j\left(u_0,\ldots,u_{j-1}\right) . \tag{4.5.32}$$

The resonances are at $j = -1$, 0, 3, 4. It can be checked that resonances at $j = -1$ and $j = 0$ correspond to the fact that ψ and u_0 are arbitrary and all the compatibility conditions are satisfied. That a logarithmic term should appear in the expansion may be understood by a similarity reduction. Set $u(x,t) = f(z) = f(x - Ct)$. The $f(z)$ satisfies an ODE that is solved by an elliptic function. And in the neighborhood of a singularity, z_0 has an expansion

$$f(z) = 2\ln(z - z_0) + \sum_{n=0}^{\infty} a_n (z - z_0)^n . \tag{4.5.33}$$

Next consider an exceptional example due to Clarkson. He considered the equation

$$u_t^2 = 2uu_x^2 - \left(1 + u^2\right)u_{xx} . \tag{4.5.34}$$

Suppose we set

$$u(x,t) = \phi^\alpha \sum_{j=0}^{\infty} u_j(t)\phi^j \tag{4.5.35}$$

with
$$\phi(x,t) = x + \psi(t) = 0 .$$

The leading order analysis indicates that $\alpha = -1$. The recursion relation turns out to be

$$j(j+1)u_0^2 u_j = F_j\left(u_0, u, \ldots, u_{j-1}, u_0', \ldots, u_{j-1}', \psi\right) .$$

It may be noted that u_0 remains arbitrary from leading order coefficient matching. One can infer that there are resonances at

$$j = 0, -1 ,$$

corresponding to the fact that u_0 and ψ are arbitrary.

We, therefore, have a second order equation with two arbitrary integration constants and two resonances, so by our usual definition it is completely integrable.

Suppose now we seek a traveling wave solution

$$u(xt) = \omega(x - ct) = \omega(z) \ ,$$

whence $\omega(z)$ satisfies

$$c^2 \left(\frac{d\omega}{dz} \right)^2 = 2\omega \left(\frac{d\omega}{dz} \right)^2 - \left(1 + \omega^2 \right) \frac{d^2\omega}{dz^2} \ ,$$

having a general solution

$$\omega(z) = \tan\left[c^{-2} \ln(AZ + B) \right] \tag{4.5.36}$$

A, B arbitrary constant. This solution has a movable essential singularity at $z = -B/A$ and therefore is not of Painleve type. We should infer that the equation from which we obtained equation (4.5.36) is not completely integrable, so where is the difficulty? This is due to the fact that the criterion set up by Weiss, Tabor, and Carnavele (based upon the algorithm of Ablowitz, Ramani, and Segur) is only necessary, not sufficient.

Now set $u = \tan W$ in (4.5.34), so W satisfies

$$W_t^2 + W_{xx} = 0 \ .$$

If $\omega = W_t$, then

$$2\omega_t\omega + \omega_{xx} = 0 \ .$$

To make a Painleve test, set

$$\omega(x,t) = \phi^\beta \sum_{j=0}^{\infty} \omega_j x, t\phi^j \ . \tag{4.5.37}$$

Lead order analysis leads to $\beta = -1$. The various coefficients of ϕ yield

$$\phi^{-3} : \omega_0 - \omega_0^2 \psi' = 0$$

$$\phi^{-2} : 2\omega_0\omega_1 - 2\omega_0^2\omega_0' = 0$$

$$\phi^{j-3} : (j-1)(j-2)\omega_j + 2\sum_{k=0}^{j-1} \omega_k'\omega_{j-k-1}$$

$$+(j-2)\psi' \sum_{k=0}^{j} \omega_k\omega_{j-k} = 0 \quad \text{for} \quad j \geq 2$$

or

$$(j+1)(j-2)\omega_j = (2-j)\psi' \sum_{k=1}^{j-1} \omega_k\omega_{j-k} - 2\sum_{k=0}^{j-1} \omega_k'\omega_{j-k-1} \qquad (4.5.38)$$

for $j \geq 2$. Therefore, the resonances are

$$j = -1, 2 .$$

The resonance $j = -1$ corresponds to the arbitrariness of ψ. But the compatibility condition for $j = 2$ is

$$\frac{d^2\omega_0^3}{dt^2} = 0 . \qquad (4.5.39)$$

Equations (4.5.38) and (4.5.39) are identically satisfied if and only if

$$\frac{d^2}{dt^2}\left\{\left(\frac{d\psi}{dt}\right)^{-3}\right\} = 0 ,$$

which may not be true for an arbitrary ψ. So ω does not have an expansion of the form (4.5.37). Terms of the form

$$[\omega_2(t) + v_2(t)\ln\phi]\phi$$

are required at $j = 2$. And at higher order, higher powers of $\ln\phi$ are required.

Therefore, equation (4.5.34) does not possess the Painleve property.

Conclusion

This chapter cites some important examples where one can visualize some deviations that may be needed in conducting the Painleve analysis. The source of difficulty is always the logarithmic terms. The present discussion points to the fact that as yet, the methodology of the analysis is not totally foolproof.

Dispersive Water-Wave equation[101]

In the concluding part of this section we consider a very important set of equations that has been widely discussed from various viewpoints. This is the Dispersive Water Wave (DWW) equation, exhaustively analyzed for its tri-Hamiltonian structure by Kupershmidt through the use of pseudodifferential operators. Before that, this set of equations was solved by inverse scattering transform by Kaup, Jaulent, and Jean. The interesting thing about the DWW set is that if one applies the ARS conjecture to it, one arrives at a third-order nonlinear equation prototype that was studied by Bureau.

The equation pair is written as

$$u_t = u_{xx} + h_x + uu_x \qquad (4.5.40)$$

$$h_t = -h_{xx} + hu_x + h_x u . \qquad (4.5.41)$$

To apply the ARS conjecture, we follow Bluman and Cole for determining the form of the similarity solution. This turns out to be

$$h = \frac{1}{t} f\left(x^2/t\right), \quad u = \frac{1}{x} g\left(x^2/t\right) . \qquad (4.5.42)$$

If we set $z = x^2/t$, then equations (4.5.41) and (4.5.42) reduce to

$$2f' = -1/z^2 \left(2g - g^2\right) + 2/z(g' - gg') - g' - 4g''$$
$$f + zf' + 2f' + 4f''z - fg / z + 2(fg)' = 0 ,$$

where

$$f'' = df^2/d^2 z , \quad g' = dg / dz .$$

On elimination of f, we get

$$16z^3 g''' = g\left(z^2 + 12\right) - 3g^3 - zg^2 + 6zg^2 g' + gg'\left(4z + 6z^2\right)$$
$$+ g'\left(-16z + z^2 - 2Az^2\right) + zA - z^2 A . \qquad (4.5.42a)$$

This equation cannot be integrated further, so it does not reduce to a second-order nODE. Although nODEs of only second order have in general been classified and analyzed in the light of Painleve analysis, one can nevertheless follow the same methodology to study any higher order nonlinear equation. Incidentally, it may be mentioned that such a study was done by Bureau and Artynor. To perform the singular point analysis, we assume

$$g \approx a_0(z - z_0)^p \ .$$

When we match $16zg'''$ and $6zg^2g'$, we get $p = -1$. Keep in mind that for the matching, each z-dependent coefficient in the neighborhood of z_0 is expanded as

$$z^3 = (z - z_0 + z_0)^3$$
$$= (z - z_0)^3 + 3z_0^2(z - z_0) + 3z_0(z - z_0)^2 + z_0^3 \ .$$

Then equation (4.5.42a) reads

$$16\Big[(z - z_0)^3 + 3(z - z_0)^2 z_0 + 3(z - z_0)z_0^2 + z_0^3\Big]g''$$
$$= g\Big[(z - z_0)^2 + 2z_0(z - z_0) + z_0^2 + 12\Big] - 3g^3 - \Big[(z - z_0) + z_0\Big]g^2$$
$$+ 6g'g^2\Big[(z - z_0) + z_0\Big] + 4gg'\Big[(z - z_0) + z_0\Big] + 6gg'\Big[(z - z_0)^2 + 2z_0(z - z_0) + z_0^2\Big]$$
$$- g'\Big[-16(z - z_0) - 16z_0 + (z - z_0)^2 + 2z_0(z - z_0) + z_0^2\Big] \ , \qquad (4.5.43)$$

where we have set the integration constant $A = 0$. To determine the resonances, we set

$$g \approx a_0(z - z_0)^{-1} + b_0(z - z_0)^{p-1} \ ,$$

so that $a_0 = 4z_0$ and p is determined by $(p+1)(p-4)(p-3) = 0$. Therefore, we have $p = -1, 4, 3$. It is not difficult to proceed further and check the arbitrariness of the expansions at these resonances, so for the third-order equation we get three resonances and three arbitrary constants, certainly conforming to the Cauchy-Kowalevsky criterion.

Another reduction that follows from the translation invariance is

$$h = h(x - vt) \ , \quad u = u(x - vt) \ , \quad \xi = x - vt \ ,$$

whence we get

$$-vu' = u'' + h' + uu'$$
$$-vh' = -h'' + (uh)' \ .$$

(4.5.44)

And if one eliminates h, one gets

$$\frac{d^2u}{d\xi^2} - v^2u \cdot + \frac{3}{2}u^2 - \frac{1}{2}u^3 = 0 \ .$$

(4.5.45)

This equation is similar to the Painleve equation and can be solved by elliptic functions. We observed that two reductions of the DWW equations satisfy all the conditions of the Painleve test. It will be interesting to proceed with the test in the sense of Weiss, Tabor, and Carnavale. We set

$$u = \sum_{j=0}^{\infty} a_j(t)\phi^{j+\alpha}(x,t)$$

$$h = \sum_{j=0}^{\infty} b_j(t)\phi^{j+\beta}(x,t) \ .$$

For the determination of the leading exponents, we set

$$u \approx a_0(t)\phi^{\alpha}(x,t)$$
$$h \approx b_0(t)\phi^{\beta}(x,t) \ .$$

Two cases then arise:

(i) $\alpha = -1, \ \beta = -2$, for which

$$a_0 = -2 , \quad b_0 = -4 \ .$$

(ii) $\alpha = -1, \ \beta = 1$ or 2, for which

$$a_0 = 2 , \quad b_0 \text{ not determined.}$$

Therefore, we have

$$u = \sum_{j=0}^{\infty} a_j(t)\phi^{j-2}(x,t)$$

$$h = \sum_{j=0}^{\infty} b_j(t)\phi^{j-2}(x,t) .$$

The recursion relations obtained are

$$a_{j-2,t} + a_{j-1}(j-2)\phi_t = (j-1)(j-2)a_j + (j-2)b_j + \sum_{k=0} a_{j-k}a_k(k-1) \qquad (4.5.46)$$

$$b_{j-2,t} + (j-3)b_{j-1}\phi_t + (j-2)(j-3)b_j = \sum_{k=0} a_{j-k}b_k(j-3) . \qquad (4.5.47)$$

The secular determinant leads to

$$(\gamma+1)(\gamma-2)(\gamma-3)(\gamma-4) = 0 .$$

Therefore, we get resonances at $\gamma = -1, 2, 3, 4$. The resonances at $\gamma = -1$ correspond to the arbitrariness of ϕ. Now we check the coefficients of the expansion from equations (4.5.46) and (4.5.47).

For $j = 0$, $b_0 = -4$, $a_0 = -2$.

For $j = 1$, $a_1 = \phi_t$, $b_1 = 0$.

For $j = 2$, $a_{0t} = 0$ (identically satisfied)

$\qquad\qquad\qquad b_2 = -2a_2$.

For $j = 3$, $b_3 = \phi_{tt}$

$\qquad\qquad\qquad b_{1t} = 0$.

For $j = 4$, $a_{2t} - a_2^2 = 2b_4 + 2a_4$

$\qquad\qquad\qquad b_{2t} + a_2b_2 + 4a_4 + 2b_4 = 0$.

These are sufficient to guarantee the arbitrariness of the coefficients, and we may conclude that DWW is an integrable system.

In the second case, we set

$$u \approx a_0\phi^{-1} + a_1\phi^{p-1}$$

$$h \approx b_0\phi + b_1\phi^{p+1} .$$

After substituting in (4.5.40) and (4.5.41) and equating coefficients of ϕ^{p-1}, we obtain

$$a_1 b_0 p + b_1 p(1-p) = 0$$

$$a(p-1)(p-2) + 2(p-2) = 0 .$$

We get

$$\det \begin{vmatrix} b_0 p & p(1-p) \\ (p-1)(p-2)+2(p-2) & 0 \end{vmatrix} = 0 .$$

Solution of this equation gives us four resonances at $p = -0, -1, 1, 2$. Arbitrariness of the coefficients can be checked as before. Now consider the other possibility of $\beta = 2$. Here,

$$u \cong a_0\phi^{-1} + a_1\phi^{p-1}$$

$$h \cong b_0\phi^2 + b_1\phi^{p+2} ,$$

leading to the resonance positions at $p = 0, -1, -1, 2$. A double resonance at $p = -1$ causes the loss of one arbitrary function. In all of the computations above, we have assumed that the coefficients are functions of 't' only and the singularity manifold ϕ is simplified as per Kruskal prescription. It may be possible to proceed in the full generality of the WTC expansion and get more interesting results. For example, if we set

$$u = \sum_{j=0}^{\infty} a_j(x,t)\phi^{j-1}(x,t)$$

$$(4.5.48)$$

$$h = \sum_{j=0}^{\infty} b_j(x,t)\phi^{j-2}(x,t) ,$$

the recursion relations so obtained are as follows:

$$a_{j-2,t} + (j-2)a_{j-1}\phi_t$$
$$= a_{j-2,xx} + 2(j-2)a_{j-1,x}\phi_x + (j-2)a_{j-1}\phi_{xx}$$
$$+ \sum_{k=0} a_{j-k}a_{kx} + \sum a_{j-k}a_k(k-1)\phi_x + (j-1)(j-2)\,a_j\phi_x^2$$
$$+ b_{j-1,x} + (j-2)b_j\phi_x$$

$$b_{j-2,t} + (j-3)b_{j-1}\phi_t + b_{j-2xx} + (j-2)(j-3)b_j\phi_x^2 + (j-3)b_{j-1}\phi_{xx} + 2(j-3)b_{j-1,x}\phi_x$$
$$= \sum_{k=0} a_{j-k-1,x}b_x + \sum_{k=0} a_{j-k-1}b_{xx} + \sum_{k=0} a_{k-k}b_k(j-3)\phi_x \; . \tag{4.5.49}$$

Resonance positions are determined as before to be at $\gamma = -1, 2, 3, 4$. For the coefficients, we find that

$$j = 0, \qquad a_0 = -2\phi_x$$
$$b_0 = -4\phi_x^2$$
$$j = 1, \qquad a_1\phi_x = \phi_t + \phi_{xx}$$
$$b_1 = 4\phi_{xx} \; .$$

Before proceeding to higher values of j, we truncate the series for u and h at the constant level as per WTC prescription, obtaining

$$u = a_0\phi^{-1} + a_1$$
$$h = b_0\phi^{-2} + b_1\phi^{-1} + b_2 \tag{4.5.50}$$

by setting $a_i = 0$ for $i \geq 2$, $b_j = 0$ for $j \geq 3$. Then for $j = 2$, we must have

$$\phi_x^2 b_2 = -2\phi_x\phi_{xt} + 2\phi_t\phi_{xx} + 2\phi_{xx}^2 - 2\phi_x\phi_{xxx} \; .$$

The other condition at $j = 2$ is identically satisfied.

For

$$j = 3, \qquad b_{1t} + b_{1xx} = a_{1x}b_1 + a_{0x}b_2 + a_1b_{1x} + a_0b_{2x}$$
$$a_{1t} = a_{1xx} + b_{2x} + a_1a_{1x} \; .$$

Lastly, for

$$j = 4, \qquad b_{2_t} = -b_{2xx} + (a_1 b_2)_x \ .$$

It is easily observed that for the consistency of these conditions, one must have

$$\phi_{tx} + \phi_{xxx} = 0$$

or

$$\phi_t + \phi_{xx} = C(t) \ , \tag{4.5.51}$$

which is nothing but a diffusion-type equation. If we consider the special case of $C = 0$, then the most general solution of (4.5.51) can be written as

$$\phi = F(t)\exp\!\left(x^2/4t\right) \ .$$

And if we apply equation (4.5.50) by starting with the trivial solution $a_1 = b_2 = 0$, we obtain

$$u = (\log \phi)_x = x\,/\,2t$$

$$h = 4\frac{\partial}{\partial x}\frac{\phi_x}{\phi} = 2/t \ ,$$

which is also a set of solutions of the DWW equation. Therefore, by the WTC approach, we have constructed a Backlund transformation for the DWW system. In the previous discussions, we have shown how a single set of equations can be treated by different variants of the Painleve analysis, and in case the requisite information may be extracted.

Chapter 5

Miscellaneous ideas in relation to Painleve analysis

5.1 Geometry and Painleve analysis

Previous discussions have explained how Painleve analysis can be effectively used to study nonlinear partial differential equations (nPDEs) and various ordinary dynamical systems. This approach is mainly analytical in nature, where the main emphasis is on the solution of the nonlinear system. On the other hand, it has been observed that integrability places strong restrictions on the geometry of the flows of the dynamical system. In a very important paper, Ercolani and Siggia[69] tried to analyze the geometric aspect of the Painleve test, which explored these restrictions and studied the reconstruction of the complete phase space. This process is called compactification. The use of flows to construct and complete a phase space is very much in use in various geometric contexts.

Consider the Riccati equation,

$$\dot{x} = a_2 x^2 + a_1 x + a_0 ,$$

(5.1.1)

with $a_i(t)$ entire functions of t. The dependent variable x is analytic whenever it exists and has only first-order poles. Set $x_1 = 1/x$ so that

$$-\dot{x}_1 = a_2 + a_1 x_1 + a_0 x_1^2 ,$$

(5.1.2)

that is x_1 also obeys a Riccati equation. The original coordinate domain $\{x \varepsilon C\}$ augments to

$$M = \{x \varepsilon C\} U \{x_1 = 0\} .$$

$x_1 = 0$ is the point at infinity that belongs to the open coordinate domain, $x_1 \varepsilon C$. These two neighborhoods, together with the transition function $\bar{x}_1 = 1/x$, define a manifold M where the set x and x_1 do not vanish. The method of Ercolani and Siggia associates coordinate neighborhood to each stratum obtained through Painleve expansion and builds up a transition function between this neighborhood and the original coordinates. The approach is that of a canonical transformation that is generated as a solution of the Hamilton-Jacobi equation of the dynamical system.

Augmented phase space

The above simple procedure for the case of the Riccati equation is called the method of adding points at infinity on the phase space. An autonomous system of ordinary differential equation (ODE) is written as

$$\dot{\bar{x}} = F(\bar{x}) , \quad x \varepsilon C^m ,$$

with F entire, analytic on C^m. Now the above system has the Painleve property if the solution

$$\bar{x} = \bar{x}(t - t_0, x_0)$$

has singularities that are at worst poles. From the standard theory of ODE, it follows that if u is a relatively compact subset of C^m, then there exists an open disc $\Delta \subset C$ such that

$$\bar{x}_{t-t_0}(\bar{x}_0) = \bar{x}(t - t_0, x_0) , \quad \Delta \times U \to C^m ,$$

so that $\bar{x}_{t-t_0}$ is a family of holomorphic maps on U parameterized by $t - t_0 \varepsilon \Delta$. It can then be shown that the phase space can be changed in such a fashion that the flow exists for all time, so that $\Delta = C$.

The conditions for the augmented phase space, M, are as follows:

1. C^m is an open dense complex submanifold of M.

2. $M - C^m$ is a finite function of irreducible analytic hypersurfaces of M.

3. The analytic flow x has the property that $\bar{x}: C \times M \to M$, which extends $\Delta \times U \to C^m$.

4. Two orbits $U_{t \in C} \cdot \bar{x}_{t-t_0}(x_0)$ and $U_{t \in C} \cdot \bar{x}_{t-t_0}(x_0')$ either coincide or disjoint.

5. If $\bar{M}$ is any complex manifold satisfying the above properties, then $M \subset \bar{M}$ is a complex submanifold.

The method of Ercolani and Siggia proposes to construct M as a union of coordinate patches that consist of $U_0 = C^m$, the original phase space, and a patch $U_i \subset C^m$ for each balance of the singularity. We denote the open cover $\{U_0, U_1, \ldots, U_m\} = V$.

For $i \neq j$, consider $U_i \cap U_j$, an open subset of U_i. To construct a complex manifold M from V, we are to define transition functions

$$\phi_{ij} : U_i \cap U_j (\subseteq U_i) \rightarrow U_i \cap U_j (\subseteq U_j),$$

so that

$$\phi_{ij}^{-1} = \phi_{ji}$$

and

$$\phi_{ij} \circ \phi_{jk} \circ \phi_{kj} = id \quad \text{on} \quad U_i \cap U_j \cap U_k.$$

It can be proved that such an M is unique.

Painleve approach to phase space

This section proceeds to the construction of the augmented phase space using the Painleve expansion. Without going into the complicated proofs, below are the four stages of construction procedure following again Ercolani and Siggia.

Procedure 1

For each balance of the singularities in the Painleve test, one develops a corresponding formal expansion of the Hamiltonian-Jacobi equation, which contains n free parameters if continued beyond a calculable order.

Procedure 2

A truncation of the above expansion then defines a canonical change of variables from $\{q, p\} = U_0$ to a patch covering the portion of infinity that corresponds to that particular balance. The patch variables are the n-free parameters and their conjugates.

Procedure 3

The Hamiltonian is rewritten in terms of these new patch variables, and the flows extend through infinity.

Procedure 4

The final step and the complicated part of the procedure is to derive transition functions among the patches. This becomes very complicated if any lower balance exists.

We proceed with a simple example to illustrate the above.

Consider the Hamiltonian H

$$2H = p^2 - 4q^3 - aq \qquad (5.1.3)$$

a is constant and (p,q) canonical. Now there are the equations of motion for this H, which are

$$\dot{q} = p$$
$$\dot{p} = -12q^2 - a \ . \qquad (5.1.4)$$

The Painleve expansion reads

$$q = \frac{1}{(t-to)^2}(1+0(t-to)+\ldots)$$

being the only principal balance. One has to add only one patch. The resonance occurs at $\rho = 6$. Because there is only one degree of freedom, we can write the solution of the Hamiltonian-Jacobi equation as

$$S = \pm\int \left(2E + aq + 4q^3\right)^{1/2} dq \ . \qquad (5.1.5)$$

The integral can be expanded for large q up to at least order $q^{-1/2} \sim (t-to)$. One has to inspect the Laurent series of p and q and use $p = \dfrac{\partial s}{\partial q}$ to infer $p \sim 2q^{3/2}$. One then sets

$$\frac{\partial s}{\partial q} = 2q^{3/2} + \frac{\partial s'}{\partial q} \qquad (5.1.6)$$

and solves

$$2q^{3/2}\frac{\partial s'}{\partial q} = \frac{a}{2}q + E - 1/2\left(\frac{\partial s'}{\partial q}\right)^2 \qquad (5.1.7)$$

recursively for s'. The first reasonable approximation $\bar{S}$ to S is

$$\tilde{S} = \pm\left(\frac{4}{5}q^{5/2} + \frac{a}{2}q^{1/2} - vq^{-1/2}\right) . \qquad (5.1.8)$$

Because S is being truncated at the order shown, E is no longer constant but becomes the variable v. We set

$$u = \partial \bar{s}/\partial v \quad \text{and} \quad p = \partial \bar{s}/\partial q ,$$

and we obtain the transition functions

$$q = u^{-2}$$
$$p = -2u^{-3} - \frac{a}{4}u - \frac{1}{2}vu^3 . \tag{5.1.9}$$

The points added to augment the manifold are $\{u = 0, \ v \in C\}$. Also,

$$dq \wedge dp = du \wedge dv .$$

The Hamiltonian in the principal patch reads

$$H(u,v) = v + \frac{a^2}{32}u^2 + \frac{a}{8}vu^4 + \frac{1}{8}v^2u^6 , \tag{5.1.10}$$

and it is easy to verify that

$$\dot{u} = \frac{\partial H}{\partial v} = 1 + O(u) .$$

One can observe that the inverse of the above map given by equation (5.1.9) appears to be multivalued around $u = 0$, but this is not the case if one remembers the domain of (5.1.9). So u can be found by solving

$$-\frac{2q}{p} = \frac{u}{1 + \frac{a}{8}u^4 + \frac{1}{4}vu^6} \tag{5.1.11}$$

recursively.

We do not proceed any further, as the construction of the manifold M becomes even more complicated in more realistic examples. The interested reader is referred to the excellent article by Ercolani and Siggia, where they show how the Painleve equation II and Henon-Heiles system can be treated.

5.2 Generation of a higher dimensional integrable system from a lower dimensional system

In a very recent communication, Sen-Yue Lou[75,84] suggested another novel application of the Painleve analysis. Though it is not yet in the form of a full-fledged theory, it will be interesting to browse through some simple examples that illustrate the idea.

Recall that a given n-dimensional N order partial differential equation written as

$$F\left(x_1, x_2, \ldots, x_n, t,\ u, u_{x_i} u_{x_i x_i}, \ldots, u_{x_i, \ldots, x_N}\right) = 0 \tag{5.2.1}$$

is said to have the Painleve property if all the movable singularities of its solution are poles. When the arbitrary singular manifold is written as

$$\phi = \phi(x_1, x_2, \ldots, x_n, t) = 0 \ ,$$

then the expansion is

$$u = \sum_{j=0}^{\infty} u_j \phi^{j+\alpha} \ , \tag{5.2.2}$$

with $(N-1)$ arbitrary functions j_j and negative integer α. If we take $u_j = 0$ $(j > -\alpha)$, we get the BT. Lou's idea was to generate some higher dimensional integrable model from such an analysis. It can be summarized as follows:

To start with, one would embed the lower dimensional model in a higher dimensional model. In other words, u will be considered not only dependent on the explicit independent variables $\{x_1, \ldots, x_n, t\}$ but also dependent on some implicit variables $\{x_{n+1}, \ldots, x_{n+m}\}$. Next, the Painleve expansion is extended to a different resummation form such that these implicit variables appear explicitly in the new expansion coefficients. The first step is easily realized from the fact that infinitely many constants of integration can be included in the solution u of the given PDE. The second step can be realized because the singular manifold ϕ is arbitrary. For instance, one can take

$$\xi = \left(\frac{\phi_{x_{n+1}}}{\phi} - \frac{\phi_{x_{n+1} x_{n+1}}}{\phi_{x_{n+1}}} \right)^{-1} \tag{5.2.3}$$

as a new expansion variable. That is, rewrite the expansion as

$$u = \sum_{j=0}^{\infty} u'_j \xi^{j+\alpha} .$$

From equation (5.2.3) it is easy to demonstrate that

$$\xi_{x_i} = P_i - P_{ix_{n+1}}\xi + (1/2)\left(P_i S + P_{ix_{n+1}x_{n+1}}\right)\xi^2 \qquad i = 0,1,2,...,n, \; x_0 = t, \qquad (5.2.4)$$

with

$$P_i = \phi_{x_i}/\phi_{x_{n+1}} \qquad (5.2.5)$$

$$S = \frac{\phi_{x_{n+1}x_{n+1}x_{n+1}}}{\phi_{x_{n+1}}} - \frac{3}{2}\left(\frac{\phi_{x_{n+1}x_{n+1}}}{\phi_{x_{n+1}}}\right)^2 .$$

P_i and S are mobius transformation invariant. The next important assertion of Lou is that if equation (5.2.1) is integrable, then the equations obtained from the conditions $u_j = 0$ for $j > -\alpha$ are integrable. A general proof of this assertion is not yet available, but we can discuss an important example from Lou.

The (2+1) dimensional KP equation is written as

$$\left(u_t - 6uu_x + u_{xxx}\right)_x + 3\sigma^2 u_{yy} = 0 . \qquad (5.7.7)$$

Denote $x_0 = t$, $x_1 = x$, $x_2 = y$, $x_3 = z$ as the basic independent coordinates. The Painleve analysis of this KP problem was done by Weiss, where from the leading order analysis it was shown that

$$\alpha = -2, \quad u'_0 = 2P_1^2 . \qquad (5.2.8)$$

The recursion relation for u'_j is

$$(j+1)(j-4)(j-5)(j-6)u'_j = f'_j\left(S, P_i, P_{ix_i}, ..., u'_0, ..., u'_{j-1}\right) , \qquad (5.2.9)$$

where f_j and S are complicated functions of the arguments shown. To mention a few u'_j,

$$u'_1 = -2P_{1x} - 2P_1 P_{1x} \qquad (5.2.10)$$

$$u'_2 = \frac{1}{6P_1^2}\left(3\sigma_2 P_2^2 + P_1\{P_0 + 4P_{1xx} + 2P_{1x}P_{1x}\}\right.$$
$$\left. + P_1^2\left(4P_{1xx} + P_{1z}^2\right) - 3P_x^2 + 4P_1^4 S + 4P_1^3 P_{1zz}\right)$$

$$u'_3 = -f'_3/24 ,$$

with a complicated expression for f_3. So the resonances are at $j = 4, 5, 6$, and all the subsequent conditions are satisfied with the required number of arbitrary functions. Now use the new conditions that

$$f'_j = 0 , \quad j = 2, 3, 7, 8, \dots .$$

For $j = 2$, $f'_2 = 0$ yields

$$\phi_x^4 \left(4\phi_x \phi_{xxx} - 3\phi_{xx}^2 + \phi_x \phi_t + 3\sigma^2 \phi_y^2 \right) + 3\phi_x^4 \phi_{zz}^2 - 6\phi_x^2 \phi_z^2 \phi_{xx} \phi_{zz} = 0 . \qquad (5.2.11)$$

Equation (5.2.11) is a new nonlinear equation that depends on four independent variables (x, y, z, t). It is now imperative to prove the integrability of this new equation. We now use the transformation

$$\phi = e^f \qquad\qquad\qquad (5.2.12)$$

to convert equation (5.2.11) into a nonhomogeneous equation,

$$f_t + 4f_{xxx} - 2f_x^3 - 3f_{xx}^2 f_x^{-1} + 3\sigma^2 f_y^2 f_x^{-1}$$
$$+ 3f_x^3 f_{zz}^2 f_z^{-4} - 6f_x f_{xx} f_{zz} f_z^{-1} = 0 . \qquad (5.2.13)$$

But the equation does not possess any algebraic poles, that is $\alpha = 0$. Therefore, one defines

$$f_x = U , \quad f_y = V , \quad f_z = W , \quad f_t = G \qquad (5.2.14)$$

in equation (5.2.13) and obtains

$$4uW^4 u_{xx} - 3W^4 u_x^2 - 2W^4 u^4 + 3u^4 W_z^2 + uGW^4$$
$$-6u^2 W^2 u_x W_x + 3\sigma^2 V^2 W^4 = 0 \qquad (5.2.15)$$

$$u_t = G_x , \quad V_t = G_y , \quad W_t = G_z .$$

We can now use the usual WTC expansion or this modified form from Lou. We set

$$U = \sum_{j=0}^{\infty} u_j \xi^{j+\alpha_1} \tag{5.2.16}$$

$$V = \sum_{j=0}^{\infty} V_j \xi^{j+\alpha_2}$$

$$W = \sum_{j=0}^{\infty} W_j \xi^{j+\alpha_3}$$

$$G = \sum_{j=0}^{\infty} G_j \xi^{j+\alpha_4} , \tag{5.2.17}$$

where ξ is given by (5.2.3), with $n = 3$, $x_0 = t$, $x_1 = x$, $x_2 = y$, $x_3 = z$ and $\alpha_1 = \alpha_2 = \alpha_3 = \alpha_4 = -1$. The leading order results give

$$u_0 = \pm P_1 , \quad V_0 = \pm P_2 , \quad W_0 = \pm P_3 , \quad G_0 = \pm P_0 .$$

Now one can prove that the resonances are at

$$j = -1, 1, 1, 1, 1.$$

The resonance condition at $j = 1$ reads

$$u_{0t} - G_{0x} + u_0 P_{0x_4} - G_0 P_{1x_4} = 0 \tag{5.2.18}$$

$$V_{0t} - G_{0y} + V_0 P_{0x_4} - G_0 P_{2x_4} = 0$$

$$W_{0t} - G_{0z} + W_0 P_{0x_4} - G_0 P_{3x_4} = 0$$

$$3u_0\left(P_1 W_0 - u_0 P_3\right)\left(W_0^2 P_{1x_4} - U_0\left(W_0 P_{3x_y} + W_{0z}\right)\right) + 2W_0^3 u_0 P_{1x}$$
$$+ W_0^2\left(W_0 P_1 - 3u_0 P_3\right)u_{0x} - 2W_0\left[\left(2u_0^2 W_0^2 - 3u_0^2 P_3^2 + 3v_0 W_0 P_1 P_3\right.\right.$$
$$\left.\left. - 2W_0^2 P_1^2\right)u_1 - u_0\left(5P_1^2 W_0 - 2W_0 u_0^2 - 3P_1 P_3 u_0\right)W_1\right] = 0 .$$

Using laborious algebraic manipulations, it can be verified that all the resonance conditions are satisfied identically, so five arbitrary coefficients ξ, U_1, V_1, W, G_1 are included in the Painleve expansion (5.2.17). Therefore, equation (5.2.11) is a $(3+1)$ dimensional nPDE obtained from the standard $(2+1)$ dimensional KP equation. The same idea can be used to construct many more higher-dimensional integrable systems from the various known integrable systems.

5.3 Conformal symmetry, Painleve test, and infinite number of symmetries

Nonlinear integrable systems are known to possess an infinite number of conserved quantities and hence an infinite number of symmetries. We have already observed that the Painleve analysis always leads to a conformal invariant equation for the singular manifold ϕ for an integrable nonlinear equation. This important fact can be used to derive various kinds to symmetry of the original nonlinear problem.

Suppose we again consider the KdV equation

$$u_t = 6uu_x + u_{xxx} = K(u) . \tag{5.3.1}$$

A symmetry of the KdV equation is defined as the solution of the linearized equation

$$\sigma_t = 6\partial_x(u\sigma) + \partial_x^3\sigma$$

$$= \lim_{\varepsilon \to 0} \frac{\partial}{\partial\varepsilon} K(u+\varepsilon\sigma) = K'\sigma \tag{5.3.2}$$

so that the infinitesimal transformation is $u \to u + \varepsilon\sigma$. In the course of the Painleve analysis, one comes across the equation

$$\phi_t = \{\phi; x\}\phi_x + 6\lambda\phi_x , \tag{5.3.3}$$

obtained as the consistency of the truncated conditions. The corresponding relation between u and ϕ is

$$u = \lambda - 1/2\left(\frac{\phi_{xx}}{\phi_x}\right)_x - 1/4\left(\frac{\phi_{xx}}{\phi_x}\right)^2 . \tag{5.3.4}$$

Now consider a symmetry transformation of equation (5.3.1) $\phi \to \phi + \varepsilon\eta$, whence it follows from relation (5.3.2) that σ and η are related as

$$\sigma = -\frac{1}{2}\partial_x\left(\frac{1}{\phi_x}\partial_x^2 - \frac{\phi_{xx}}{\phi_x^2}\partial_x\right)\eta - 1/2\left(\frac{\phi_{xx}}{\phi_x}\right)\left(\frac{1}{\phi_x}\partial_x^2 - \frac{\phi_{xx}}{\phi_x^2}\right)\eta . \tag{5.3.5}$$

It is now important to notice that equation (5.3.3) is conformally invariant. That is, a symmetry transformation is

$$\phi \to \frac{a+b\phi}{c+d\phi}\,(ad \neq bc)\,.$$

Consider the special case. $a = 0$, $b = c = 1$, and $d = \varepsilon$, whence the above transformation yields for infinitesimal ε :

$$\phi \to \phi - \varepsilon\phi^2 \,. \tag{5.3.6}$$

A solution for η is $-\phi^2$, so using the relation (5.3.5), we get

$$\sigma = 2\phi_{xx}\,.$$

On the other hand, it is known that the substitution

$$\phi_x = \frac{1}{2}\psi^2 \tag{5.3.7}$$

leads to the Schrodinger problem for the KdV equation

$$\psi_{xx} + u\psi = \lambda\psi \,. \tag{5.3.8}$$

Therefore, one gets

$$\sigma = 2\phi_{xx} = \partial_x\psi^2 \,. \tag{5.3.9}$$

Starting from the single nonlocal symmetry $\sigma = \partial_x\psi^2 = K_0^1(\lambda)$, we can generate an infinite number of symmetries of the KdV equation if we expand σ as follows:

$$K_0^1(\lambda) = \sum_{n=0}^{\infty}\frac{1}{n!}\left(\frac{\partial^n}{\partial\lambda^n}K_0^1(\lambda)\bigg|_{\lambda=0}\right)\cdot\lambda^n \,. \tag{5.3.10}$$

To obtain these, we formally solve equation (5.3.8). That is,

$$\psi(\lambda) = \sum_{k=0}^{\infty}\left(\partial_x^2 + u\right)^{-1}\psi_0\lambda^k$$

$$= \sum_{k=0}^{\infty}\left(\psi_0\partial_x^{-1}\psi_0^{-2}\partial_x^{-1}\psi_0\right)^k\psi_0\lambda^k \tag{5.3.11}$$

when

$$\psi(\lambda = 0) = \psi_0.$$

One can therefore obtain from equation (5.3.11)

$$K_n^1 = 2\sum_{k=0}^{n} \left(\left(\psi_0 \partial_x^{-1} \psi_0^{-2} \partial_x^{-1} \psi_0\right)^k \psi_0 \right) \left(\psi_0 \partial_x^{-1} \psi_0^{-2} \partial_x^{-1} \psi_0\right)^{n-k} \left(\psi_0^2\right)_x$$

$$= 2^{2n} \left(\partial_x \psi_0^2 \partial_x^{-1} \psi_0^{-2} \partial_x^{-1} \psi_0^{-2} \partial_x^{-1} \psi_0^2\right)^n \left(\psi_0^2\right)_x$$

$$= 2^{2n} \phi^{-n} \left(\psi_0^2\right)_x , \qquad\qquad\qquad (5.3.12)$$

ϕ being the recursion operator for the KdV problem. One may also note that if ψ is a solution of (5.3.8), then

$$C_1\psi + C_2\psi \cdot \partial_x^{-1}\psi^2 \qquad\qquad\qquad (5.3.13)$$

is also a solution of the Schrodinger equation. Proceeding with this, one may obtain the other two sets of nonlocal symmetries:

$$K_n^{(2)} = 2^{2n} \phi^{1-n} \left(\psi_0^2 \partial_x^{-1} \psi_0^{-2} \partial_x^{-1} \psi_0^{-2}\right)_x$$

$$K_n^{(3)} = 2^{2n} \phi^{-n} \left(\psi_0^2 \partial_x^{-1} \psi_0^{-2}\right)_x . \qquad\qquad (5.3.14)$$

By making other choices of (a, b, c, d) in equation (5.3.6), it is possible to obtain many more symmetries of the KdV equation. So by starting from the invariance properties of the singular manifold equation, it is possible to obtain nonlocal symmetries of the KdV equation.

The same technique may be applied to other nonlinear systems, and it is quite convenient and useful for the study of nonlocal symmetries.

5.4 Painleve analysis and Painleve equations

One of the earliest analyses of the concept of complete integrability started with the observation that nonlinear PDEs often reduce to nonlinear ODEs of the Painleve class if they are completely integrable. This was the starting point of the seminal paper by Ablowitz, Ramani, and Segur.[36b,74] This is known as the ARS conjecture. We have already seen examples of such reduction in various sections of the previous chapters. But a recent observation of Kudryashov is worth mentioning. He observed that if the Lie symmetry reduction is applied to the whole hierarchy of an integrable system instead of only to the basic one, then some new higher-order nonlinear ODEs result in Painleve equations of higher order.

Start with the KdV equation

$$q_t + 3qq_x + q_{xxx} = 0 \tag{5.4.1}$$

and set

$$q = F(\theta) - \lambda t, \qquad \theta = x + \frac{3}{2}\lambda t^2.$$

Then it is easy to observe that $F(\theta)$ obeys

$$F_{\theta\theta} + \frac{3}{2}F^2 - \lambda\theta + K_1 = 0. \tag{5.4.2}$$

But for the modified KdV case,

$$u_t - \frac{3}{2}u^2 u_x + u_{xxx} = 0. \tag{5.4.3}$$

A solution has the form

$$u = \frac{1}{(3t)^{1/3}} f(\xi), \qquad \xi = \frac{x}{(3t)^{2/3}}. \tag{5.4.5}$$

This form is suggested by the Lie point symmetry of the MKdV equation. It is then clear that $f(\xi)$ follows

$$f_{\xi\xi} - \frac{7}{2}f^3 - \xi f + K_2 = 0. \tag{5.4.6}$$

In the above equations, K_1 and K_2 are integration constants. Equation (5.4.2) is the first Painleve, and equation (5.4.6) is known as the second Painleve ODE.

Next consider the hierarchy of the KdV equation

$$\omega_t + \frac{\partial}{\partial x} B^n(\omega) = 0, \tag{5.4.7}$$

where B^n follows the recursion relation

$$\frac{\partial}{\partial x} B^{n+1}(\omega) = B^n_{xxx} + 2\omega B^n_x + \omega_x B^n \tag{5.4.8}$$

$$B^0 = 1, \qquad B' = \omega .$$

On the other hand, from the Painleve analysis it is known that the singular manifold equation corresponding to the hierarchy is

$$Z_t + Z_x B^n(\{z, x\}) = 0 , \tag{5.4.9}$$

where

$$\{Z, x\} = \frac{Z_{xxx}}{Z_x} - \frac{3}{2} \frac{Z_{xx}^2}{Z_x^2} .$$

If one now applies the similarity reduction procedure to this equation, one obtains the higher-order Painleve equation. For example, for $n = 2$ one gets

$$F_{zzzz} + 5FF_{zz} + 5/2F_z^2 + 5/2F^3 - \lambda z = 0 , \tag{5.4.10}$$

which is called the first Painleve equation of fourth order, where z is the nw similarity variable. Similarly, in the case of the MKdV system, the hierarchy is written as

$$u_t + \frac{\partial}{\partial x}\left[\left(\frac{\partial}{\partial x} + u\right) B^n \left(u_x - 1/2u^2\right)\right] = 0 , \tag{5.4.11}$$

where B is again given by equation (5.4.8). In this case, the similarity variables can be invoked if we set

$$u = [(2n+1)t]^m f(\xi) \tag{5.4.12}$$

$$\xi = x[(2n+1)t]^m \qquad m = -1/(2n+1) .$$

For the case $n = 2$, this leads to

$$f_{\xi\xi\xi\xi} - \frac{5}{2}f^2 f_{\xi\xi} - \frac{5}{2} \cdot ff_\xi^2 + \frac{3}{8}f^5 - \xi f + K_2 = 0 , \tag{5.4.13}$$

which is the second Painleve equation of fourth order. On the other hand, one can have an identity connecting the form of the hierarchy in terms of the singular manifold (of Painleve analysis) and its usual representation in the following form:

$$
u_t + \frac{\partial}{\partial x}\left[\left(\frac{\partial}{\partial x}+u\right)B^n\left(u_x - 1/2u^2\right)\right]
$$

$$
= \left(\frac{1}{z_x}\frac{\partial}{\partial x} - \frac{2}{z}\right)\left[z_t + z_n B^n(\{z;x\})\right].
\tag{5.4.14}
$$

Using the similarity variables (ξ), one gets

$$
\left(\frac{d}{d\xi}+f\right)d^n\left(f_\xi - 1/2f^2\right)-\xi f - 1
$$

$$
= \left(\frac{1}{z\xi}\frac{d}{d\xi}z_\xi - \frac{2z_\xi}{z}\right)\left[d^n(\{z;\xi\})-\xi\right],
\tag{5.4.15}
$$

which is very important because the left side contains the second Painleve equation of $(2n)$th order and the right side represents the first Painleve equation of $(2n-2)$th order. Further properties of such higher-order Painleve equations are yet to be studied and analyzed, and the topic is very important for future research.

5.5 The Ablowitz, Ramani, Segur approach

Due to the pioneering importance of the paper by Ablowitz, Ramani, and Segur, a short discussion of their idea is shown below, which was the only method for the analysis of nonlinear partial differential system prior to the work of Weiss, Tabor, and Carnavale.

In their original approach Ablowitz, Ramani, and Segur considered the various symmetry reductions of a given nPDE and then performed the Painleve analysis.

The basic idea of the reduction comes from the linear integral equation used in the inverse scattering analysis of a system. The integral equation is written as

$$
K(x,y,t) = F(x+y;t) + \int_x^\infty K(x,z;t)N(x,z,y;t)dz,
\tag{5.5.1}
$$

where N is given in terms of F in various ways:

1. $N(x,z,y,t) = F(z+y,t)$
\hfill (5.5.2)

2. $N(x,z,y;t) = \pm \int_{x}^{\infty} F(z+u;t)F(u+y;t)du$ (5.5.3)

3. $N(x,z,y,t) = \pm \int_{x}^{\infty} \overline{F}(z+u;t)F(u+y;t)du$ (5.5.4)

$\overline{F}(x,t) = [F(x^{*};t^{*})]^{*}$ (5.5.5)

* means complex conjugation.

4. $N(x,z,y,t) = \pm \int_{x}^{\infty} \partial_{z} F(z+u,t)F(u+y,t)du$ (5.5.6)

5. $N(x,z,y,t) = \int_{x}^{\infty} \int_{x}^{\infty} F(z+u,t)F(u+v,t)F(v+y,t)du\,dv$, (5.5.7)

where F satisfies a linear PDE and the nonlinear field variable u is given as

$u = K(x,x,t)$

or (5.5.8)

$u = \dfrac{d}{dx} K(x,x,t)$.

Of the various types of nonlinear equations there are many that admit self-similar reductions. In such a case, one can ascertain that

$F(x+y;t) = \phi(x+y;t)F_{0}[\psi(xt)+\psi(yt)]$

(5.5.9)

$K(xyt) = \phi(x+y;t)K[\psi(xt),\psi(yt)]$.

ϕ and ψ are known functions, and this ansatz reduces the linear PDE of F to an ODE of F_{0}. The integral equation then has a solution that is the self-similar solution of the nonlinear PDE.

The basic ingredients of this approach are the following:

1.   The basic integral equation connecting K and F.

2.   The two differential equations (linear) satisfied by F,

$$L_i F = 0 \ ,$$

which can be related to the space and time dependence of the Lax eigenfunction.

3.   In the ARS formulation,

$$L_1 = \partial_x - \partial_y \ , \tag{5.5.10}$$

and K is connected to F by

$$(1 - Ax)K = F \ , \tag{5.5.11}$$

a short form of the aforementioned integral equation.

4.   If we apply L_i to this equation, then

$$L_i(1 - Ax)K = 0 \quad \text{for} \quad i = 1, 2 \tag{5.5.12}$$

or

$$(1 - Ax)L_i K = R_i \tag{5.5.13}$$

where Ri is generated by moving α_i across $(1 - Ax)$. But in many situations it happens that

$$R_i = (1 - Ax)M_i(K) \ . \tag{5.5.14}$$

$M_i(K)$ are nonlinear functionals of K, so one gets

$$(1 - Ax)[\alpha_i K - M_i(K)] = 0 \ . \tag{5.5.15}$$

But because $(1 - Ax)$ invertible, K must be a solution of the nonlinear equation

$$L_i K - M_i(K) = 0 \ . \tag{5.5.16}$$

Therefore, we have a correspondence that every solution of the linear integral equation is also a solution of the nonlinear equation. The form of $L_2 F = 0$ may be of various forms:

1. $\quad i\partial_t F = Q(i\partial_x)F$ (5.5.17)

   is of evolution type. Q is a polynomial.

2. Linear ODEs for F are obtained by reducing with suitable ansatz:

$$F(x+y,t) = \tilde{F}(x+y-2vt)$$
$$K(x,y,t) = \tilde{K}(x-vt, y-vt) \ .$$
(5.5.18)

3. If $Q(i\partial_x) = (i\partial_x)^p$, then a scale invariance can be invoked to set

$$F(x,t) = t^{-1/p} f\left(xt^{-1/p}\right)$$

or
(5.5.19)
$$F(x,t) = x^{-1/2} f\left(xt^{-1/p}\right) .$$

Because it is difficult to enumerate the various possibilities in the general situation, it is convenient to illustrate in some particular case as was done by Ablowitz, Ramani, and Segur.

The modified KdV case
In this case, the operator $L_1 = (\partial_x - \partial_y)$ and the solution of $L_1 F = 0$ is $F = F((x+y)/2)$. Therefore, the integral equation for K becomes

$$K(x,y) = F\left(\frac{x+y}{2}\right) + \frac{\sigma}{4} \int_x^\infty \int K(xz) \cdot F\left(\frac{z+u}{2}\right) F\left(\frac{u+y}{2}\right) dz du(\alpha) .$$
(5.5.20)

On the other hand,

$$L_2 = \partial_t + \partial_x^2 \quad \text{and} \quad L_2 F = 0 .$$

The methodology of AKS is to shift origin so that the boundary terms are avoided,

$$K(x,y)= F\left(\frac{x+y}{2}\right)+\sigma/4\int\!\!\int_{0}^{\infty}K(x,x+s)F\left(\frac{2x+s+s}{2}\right)F\left(\frac{x+s+y}{2}\right)dsds \quad (5.5.21)$$

or

$$\left[(I-\sigma A_x)K\right](x,y)= F\left(\frac{x+y}{2}\right).$$

Now set

$$K_2(x,z)= \int_{0}^{\infty} K(x,x+p)F\left(\frac{x+p+z}{2}\right)dp . \qquad (5.5.22)$$

It then easily follows that

$$(1-\sigma A_x)K_2(x,z)= \int_{0}^{\infty} F\left(\frac{2x+p}{2}\right)F\left(\frac{x+p+z}{2}\right)dp . \qquad (5.5.23)$$

It is now convenient to write the integral equation (5.5.21) as

$$K(x,y)= F\left(\frac{x+y}{2}\right)+\sigma/4\int_{0}^{\infty} K_2(x,x+q)F\left(\frac{x+q+y}{2}\right)dq . \qquad (5.5.24)$$

Applying the operator $\left(\partial_x-\partial_y\right)$ to this, one gets

$$\left(\partial_x-\partial_y\right)K(x,y)= \sigma/4\int_{0}^{\infty}(\partial_1+\partial_2)K_2(x,x+q)F\left(\frac{x+q+y}{2}\right)dq . \qquad (5.5.25)$$

Here ∂_1 and ∂_2 are the derivatives with respect to the first and second arguments. Similarly, if we apply $\left(\partial_x+\partial_z\right)$ to equation (5.5.22), we get

$$(\partial_x+\partial_z)K_2(xz)$$

$$= \int_{0}^{\infty}\left\{(\partial_1+\partial_2)K(x,x+p)F\left(\frac{x+p+z}{2}\right)+K(x,x+p)F'\left(\frac{x+p+z}{2}\right)\right\}dp$$

$$= \int_{0}^{\infty}[(\partial_1-\partial_2)K(x,x+p)]F\left(\frac{x+p+z}{2}\right)dp -2K(xx)F\left(\frac{x+z}{2}\right). \qquad (5.5.26)$$

Substituting (5.5.25) into this equation,

$$\left(1-\sigma A_x\right)\left(\partial_x+\partial_y\right)K_2(x,z)$$

$$=-2K(xx)F\left(\frac{x+z}{2}\right)$$

$$=-2K(xx)\left(1-\sigma A_x\right)K(xz) .$$
(5.5.27)

On the other hand, if equation (5.5.26) is used in (5.5.25), we get

$$\left(1-\sigma A_x\right)\left(\partial_x-\partial_y\right)K(x,y)\infty$$

$$=-\frac{\sigma}{2}K(x,x)\int_0^\infty F\left(\frac{2x+q}{2}\right)F\left(\frac{x+q+y}{2}\right)dq$$

$$=-\frac{\sigma}{2}K(x,x)\left(1-\sigma A_x\right)K_2(x,y) .$$
(5.5.28)

Also, A_x commutes with any multiplicative function of x, so $\left(1-\sigma A_x\right)$ is invertible. We have

$$\left(\partial_x+\partial_z\right)K_2(x,z)=-2K(x,x)K(xz)$$

$$\left(\partial_x-\partial_y\right)K(x,y)=-\frac{\sigma}{2}K(xx)K_2(x,y) .$$
(5.5.29)

If one now applies $\left(\partial_x+\partial_y\right)$ to (5.5.20), one gets, after some tedious algebra,

$$F'\left(\frac{x+y}{2}\right)=(I-\sigma Ax)\left\{\left(\partial_x+\partial_y\right)K(x,y)+\frac{\sigma}{2}K_2(x,x)K(xy)\right\} .$$

It is now important to recapitulate that F also satisfies $L_2F=0$. Again apply $L_2=\partial_t+\left(\partial_x+\partial_y\right)^3$ to (5.5.20) and obtain

$$\left\{\left(\partial_x+\partial_y\right)^3+\partial_t\right\}K(x,y)$$

$$=\sigma/4\left\{\partial_t+\left(\partial_x+\partial_y\right)^3\right\}\int\int_0^\infty K(x,x+p)F\left(\frac{2x+p+q}{2}\right)F\left(\frac{x+q+y}{2}\right)dpdq .$$
(5.5.30)

Also, from (5.5.29) we get

$$\partial_x K_2(x,x) = -2K^2(x,x) \,,$$ (5.5.31)

so that one gets

$$\left\{ \partial_t + \left(\partial_x + \partial_y\right)^3 \right\} K(x,y) = 3\sigma K(xx) K(x,y) \partial_x K(x,x)$$

$$+3\sigma K^2(x,x)\left(\partial_x + \partial_y\right) K(x,y) \,.$$ (5.5.32)

If we now define

$$q(x,t) = K(x,x,t)$$

and evaluate the above equation along $y = x$, we get

$$\partial_t q + \partial_x^3 q = 6\sigma q^2 q_x \,.$$

q satisfies the modified KdV equation, thus every solution of $L_i F = 0$ with $i = 1, 2$ that decays fast enough as $x \to \infty$ defines a solution of (5.5.20) via the linear integral equation.

Now consider the case when K and F are similar:

$$K(x,y,t) = (3t)^{-1/3} k(\xi,n)$$

$$F\left(\frac{x+y}{2},t\right) = (3t)^{-1/3} f\left(\frac{\xi+n}{2}\right),$$ (5.5.33)

where
$$\xi = x/(3t)^{1/3} \,, \qquad \eta = y/(3t)^{1/3} \,.$$

Substituting these in equation (5.5.21), we get

$$k(\xi,n) = f\left(\frac{\xi+n}{2}\right) + \sigma/4 \int\int_{}^{\infty} k(\xi,z) f\left(\frac{z+\psi}{2}\right) f\left(\frac{\psi+n}{2}\right) dz d\psi \,.$$ (5.5.34)

Substituting (5.5.23) in the equation of the same form as (5.5.21) but with shifted origin, that is $(-\alpha)$, we get

$$f'''(\xi) - f(\xi) - \xi f'(\xi) = 0 \,.$$ (5.5.35)

Integrating once

$$f''(\xi) - \xi f(\xi) = C_1 .$$

If $C_1 = 0$, then the solution that goes to 0 as $\xi \to \alpha$ is

$$f\left(\frac{\xi+n}{2}\right) + \gamma Ai\left(\frac{\xi+n}{2}\right) , \tag{5.5.36}$$

where $Q(\xi) = k(\xi, \xi)$ must be a solution of

$$Q''' - [Q + \xi Q'] = 6\sigma Q^2 Q' ,$$

which can also be integrated to

$$Q'' = \xi Q + 2\sigma Q^3 + C_2 .$$

This nonlinear ODE is the second Painleve equation. Therefore, the above discussion points to the fact that for each solution of the integral equation (5.5.20), there is one solution of the second Painleve equation. If $C_2 = 0$, then we can use (5.5.36) in the equation (5.5.34) to obtain

$$\left[1 - \sigma\gamma^2 \overline{A}_\xi\right] k(\xi, n) = \gamma Ai\left(\frac{\xi+n}{2}\right) ,$$

where

$$\overline{A}_\xi f(\eta) = 1/4 \int\int^{\infty} f(\xi) Ai\left(\frac{p+q}{2}\right) Ai\left(\frac{p+n}{2}\right) dp dq .$$

The only singularities of these solutions in the complex planes are poles, which was first studied by Painleve with his $\alpha-$ method. Here it can be proved by using the Fredholm theory of the integral equation.

The idea of reducing a nPDE to an ordinary differential system to study the analytic structure paved the way for future developments that are still used in the uncharted zone of a new and complicated nPDEs.

5.6 Painleve analysis of hierarchy

Each Lax pair not only gives rise to a single integrable partial differential equation, but it also generates a hierarchy of nonlinear systems that shares many common properties. From this point of view, it is interesting to inquire about the complete integrability of a single hierarchy rather than a single equation. Suppose we write an nPDE as

$$u_t = K[u] . \tag{5.6.1}$$

The Painleve analysis seeks a solution in the form

$$u = \phi^{-\alpha}(x,t) \sum_{j=0}^{\infty} u_j(x,t)\phi^j(x,t) , \tag{5.6.2}$$

with $\phi(x,t) = 0$ being the singular manifold. The whole process starts with three choices:

1. leading order exponent α,
2. leading order coefficient u_0,
3. dominant terms $K_d(u)$.

For each such set, there is a set of indices $R = \{\gamma_1, \gamma_2, ..., \gamma_N\}$ that will give the resonance positions where the arbitrary coefficients are introduced. So the whole family of parameters and variables can be collected in a set S_1

$$S_1 : \{\alpha, u_0, K_d(u), \beta, \{\gamma_1, ..., \gamma_N\}\} .$$

β is the weight of $K_d(u)$, when u is of weight α and ∂/∂_x of weight 1. The set $\{S_1\}$ satisfies

$$K_d[u_0\phi^{-\alpha}] = 0$$

$$\phi_{xx} = u_{0x} = 0 , \tag{5.6.3}$$

whence the resonances are determined as the root of equation

$$Q(\gamma, \alpha, v_0) = \phi^{\beta-\gamma} K_d'[u_0\phi^{-\alpha}]\phi^{\gamma-\alpha} = 0 \tag{5.6.4}$$

when

$$\phi_x - 1 = U_{0x} = 0 \ .$$

K'_d is the Frechet derivative of K_d. This section discusses how the present approach may be applied to the case of hierarchy of equation. Suppose the hierarchy is written as

$$u_{t_{n+1}} = K_{n+1}[u]$$

$$= R^n[u]K_1[u] \ ; \quad n = 0, 1, 2, \dots \ . \tag{5.6.5}$$

$R[u]$ is some recursion operator. Also note that $K_1[u] = u_x$ is the linear flow. The method for constructing the equation $Q_{n+1}(\gamma, \alpha, v_0) = 0$ for the hierarchy was given by Pickering. Here we shall follow his treatment for the analysis and derivation of $Q_{n+1}(\gamma, \alpha, u_0)$. It is not difficult to deduce that

$$K'_{n+1}\left[u_0\phi^{-\alpha}\right]\phi^{\gamma-\alpha} = 0 \tag{5.6.6}$$

for the t_{n+1} flow. Also, we get

$$K'_{n+1}\left[u_0\phi^{-\alpha}\right]\phi^{\gamma-\alpha}$$

$$= R'\left[u_0\phi^{-\alpha}\right]K_n\left[u_0\phi^{-\alpha}\right]\phi^{\gamma-\alpha}$$

$$+ R\left[u_0\phi^{-\alpha}\right]K_n'\left[u_0\phi^{-\alpha}\right]\phi^{\gamma-\alpha} \ . \tag{5.6.7}$$

Each of the terms is to be considered when $\phi_x - 1 = u_{0x} = 0$. On the other hand, if our equation hierarchy is also a family of E_n-flow, then

$$K'_n\left[u_0\phi^{-\alpha}\right]\phi^{\gamma-\alpha} = 0 \ , \quad \text{when} \quad \phi_x - 1 = 0$$

or

$$R\left[u_0\phi^{-\alpha}\right]K'_n\left[u_0\phi^{L-\alpha}\right] = 0 \ . \tag{5.6.8}$$

To proceed further, consider the special case of KdV hierarchy, which is known to be the sequence of bi-Hamiltonian flow

$$u_{t_{2n+1}} = K_{2n+1}[u]$$

$$= R^n[u]u_x$$

$$= B_0 \delta H_{n+1} = B_1 \delta H_n \qquad n = 0, 1, 2, \ldots \tag{5.6.9}$$

when

$$B_0 = \partial$$

$$B_1 = \partial^3 + 2(u\partial + \partial u)$$

and

$$R = B_1 B_0^{-1} = \partial^2 + 4u + 2u_x \partial^{-1} .$$

Now from the expression (5.6.9), it is easy to show that $\alpha = 2$ and

$$v_0 = -K(k+1)\phi_x^2 \qquad K = 1, 2, \ldots n . \tag{5.6.10}$$

It may be noted that because (5.6.9) is of weight $2n+3$ and is in the form of a conservation law, each family of t_{2n+1} flow has an index at $\gamma = 2n+2$. Furthermore, the kth family of the t_{2n+1} flow has k negative indices,

$$\gamma = -(2m+1) , \qquad 0 \le m \le K-1 , \tag{5.6.11}$$

which arise as a direct consequence of the presence of the K lower-order commuting flows,

$$u_{t_{2m+1}} = K_{2m+1}[u] \qquad 0 \le m \le K-1 .$$

(For a detailed discussion of this point, refer to Section after equation (5.6.16))

Now consider the ODE obtained

$$\delta H_{n+1} = u_{(2n)x} - H_{n+1}\left[u_{(2n-1)x}, \ldots, u\right] = 0 , \tag{5.6.11}$$

which we rewrite as the $2n$-dimensional system, $\left(U = U^0\right)$

$$u_x^{(0)} = u^{(1)}$$

$$\ldots$$

$$u_x^{(2n-2)} = u^{(2n-1)}$$

$$u_x^{(2n-1)} = H_{n+1}\left[u^{(2n-1)}, \ldots, u^{(0)}\right] . \tag{5.6.12}$$

The leading-order behavior of the original system includes a leading-order behaviour of (5.6.12), and the corresponding behavior of the two systems have the same indices. This ODE can be studied from the viewpoint of Kowalevsky's exponent. Information about the first integral of (5.6.12) are provided by the fluxes of (5.6.9), which are easily generated by the bi-Hamiltonian nature of it.

$$\partial(\delta H_{m+1})(\delta H_{n+1}) = \{H_{n+1}, H_{m+1}\}_\partial = (G_{n,m})_x \tag{5.6.13}$$

for some density $G_{n,m}$ of weight $2n + 2m + 4$, which are actually the conserved quantities. Equation (5.6.13) now implies that

$$\frac{\partial G_{n,m}}{\partial_U (2n-1)} = \frac{\partial G_{n,m}}{\partial_u (2n-1)} = \partial \delta H_{m+1} = K_{2m+1}[u] \tag{5.6.14}$$

the various types of ODEs so obtained. It was their conjecture that if all possible reductions of a given PDE conform to the Painleve criterion, then the original PDE will also be completely integrable. An application of such an idea to the case of dispersive water waves is described in Chapter 4, Section 4.5.

For $m > K$, the family (5.6.10) is a family of t_{2m+1} flow, and for these values of u_0,

$$K_{2m+1}\left[u_0 \phi^{-2}\right] = 0 . \tag{5.6.15}$$

So the leading order coefficient and the index can be fixed for the whole hierarchy. Now look at the system in a different manner. Rewrite the hierarchy as

$$q_{t_{2n+1}} = \frac{\partial}{\partial x} L^n q \quad n \geq 0 \tag{5.6.16}$$

with

$$L = -\frac{1}{4} \frac{\partial^2}{\partial x^2} - q + 1/2 \int^x dx\, q_x .$$

Substitute

$$q = \frac{u_0}{\phi^2} + \frac{u_1}{\phi} + u_2 + u_3\phi + \dots , \tag{5.6.17}$$

which yields

$$L^n q = \frac{1}{\phi^{2n+2}} \sum_{l=0}^{\infty} A_1^{(n)} \phi^l \ .$$

To determine the recursion relation of $A_1^{(n)}$, evaluate $L^n q$ via $L^{n-1}q$, that is

$$\frac{1}{\phi^{2n+2}} \sum_{l=0}^{\infty} A_1^{(n)} \phi^l = \left(-\frac{1}{4}D^2 - q + \frac{1}{2}D^{-1}q_x \right) \frac{1}{\phi^{2n}} \sum_{l=0}^{\infty} A_1^{(n-1)} \phi^l \tag{5.6.18}$$

which yields

$$\left(-\frac{1}{4}D^2 - \frac{1}{\phi^2} \sum_{l=0}^{\infty} u_1 \phi^l \right) \frac{1}{\phi^{2n}} \sum_{l=0}^{\infty} A_1^{(n-1)} \phi^l$$

$$= \frac{1}{\phi^{2n+2}} \sum_{l=0}^{\infty} \left\{ -\frac{1}{4}A_{1-2,xx}^{n-1} + 1/2(2n-1+1)A_{1-1,x}^{n-1}\phi_x \right.$$

$$+1/4(2n-\ell+1)A_{1-1}^{n-1}\phi_{xx} - 1/4(2n-1)(2n-1+1)A_1^{n-1}\phi_x$$

$$\left. -\sum_{l=0}^{\infty} A_i^{(n-1)} u_{1-i} \right\} \phi^\ell \ . \tag{5.6.19}$$

Since $q_x L^{n-1} q$ is an exact derivative, we have

$$q_x L^{n-1} q = D \left[\frac{1}{\phi^{2n+2}} \sum_{l=0}^{\infty} C_1^n \phi^l \right], \tag{5.6.20}$$

which is equivalent to

$$\frac{1}{\phi^{2n+3}} \sum_{i=0}^{\infty} \left\{ \sum_{i=0}^{1} A_i^{(n-1)} \left[u_{1-i-1,x} + (1-i-2)u_{1-i}\phi_x \right] \right\} \phi^l$$

$$= \frac{1}{\phi^{2n+3}} \sum_{l=0}^{\infty} \left[C_{1-1x}^n - (2n-1+2)\phi_x C_1^{(n)} \right] \phi^l \ , \tag{5.6.21}$$

which yields

$$C_1^n = \frac{1}{(2n-1+2)\phi_x} \left\{ -\sum_{i=0}^{\ell} A_i^{(n-1)} \left[u_{1-i-1,x} + (1-1-2)u_{1-i}\phi_x \right] + C_{1-1,x}^n \right\}. \qquad (5.6.22)$$

Therefore, equations (5.6.18) and (5.6.21) imply

$$A_1^n = -1/4(2n-1)(2n-1+1)A_1^{n-1}\phi_x^2 + 1/2(2n-1+1)A_{1-1,x}^{n-1}\phi_x$$

$$+ 1/4(2n-1+1)A_{1-1}^{n-1}\phi_{xx} - 1/4A_{1-2,xx}^{n-1} + 1/2C_1^n - \sum_{i=0}^{1} A_i^{n-1}u_{1-1}, \qquad (5.6.23)$$

which is the required recursion for A_1^n. Some explicit values of these coefficients are

$$A_0^K = (-1)^k \frac{(2k+1)}{2^k(k+1)} \prod_{i=0}^{k} \left(u_0 + i(i+1)\phi_x^2 \right)$$

$$A_1^K \left(u_0 = -2\phi_x^2 \right) = (-1)^{k+1} \frac{(2k+3)K!(2k-3)!!}{2^k} (u_1 - 2\phi_{xx})\phi_x^{2k}$$

$$A_2^K \left(u_0 = -2\phi_x^2, \ u_1 = 2\phi_{xx} \right) = 0 \qquad (5.6.24)$$

$$A_3^K \left(u_0 = -2\phi_x^2, \ u_1 = 2\phi_{xx}, \ \phi_x = \psi^2 \right)$$

$$= (-1)^k \frac{(k-1)!(2k-5)!!(2k+1)}{2^{k-1}} \psi^{4k-2} \left[\frac{\psi_{xx}}{\psi} + u \right]_x. \qquad (5.6.25)$$

Similar expressions can be obtained for other coefficients as well. The proof of these results involves laborious algebraic manipulations and is omitted here. But for the general value,

$$U_0 = -m(m+1)\phi_x^2. \qquad (5.6.26)$$

These coefficients turn out to be

$$A_{2n+1}^n = (-1)^{n+1} 2\phi_x^2 \frac{(2n+1)!!}{(2n+2)!!} \prod_{m=1}^{n} \left(u_0 + m(m+1)\phi_x^2 \right), \qquad (5.6.27)$$

and A_1^n is proportional to

$$\phi_x^{2n}\left(u_1 - m(m+1)\phi_{xx}\right) .$$

So there are n-branches of the Painleve expansion

$$q(x,t) = m(m+1)\frac{\partial^2}{\partial x^2}\ln\phi + \sum_{2}^{\infty} u_j^{(m)}\phi^{j-2} . \tag{5.6.28}$$

For each choice of m (1 to n), the sequence $\left\{u_j^{(m)}\right\}_{j=2}^{\infty}$ will be different and will have a different number of undetermined functions at the resonance positions. For the principal balance $m=1$, we will prove that there are $(2n-1)$ resonances that give rise to $(2n-1)$ undetermined functions in addition to ϕ. For an intermediate branch, $1 < m < n$, the corresponding Painleve series has $2n-m$ undetermined functions. If we take recourse to Kruskal's simplification, then these functions only depend on t. To ascertain the structure of resonances, note that C_1^n and A_1^n do not involve u_k for $k>1$. Hence we can write

$$A_k^n\left(u_0, u_1, \ldots, u_k\right)$$

$$= -1/4(2n-k)(2n-k+1)A_k^{n-1}\phi_x^2 - A_0^{n-1}u_k - A_k^{n-1}u_0$$

$$- \frac{(k-2)A_0^{n-1}}{2(2n-k+2)}u_k + \frac{u_0}{2n-k+2}A_k^{n-1} + A_k^{-n}\left(u_0, \ldots, u_{k-1}\right) ,$$

which reduces to

$$A_k^n\left(u_0 = -m(m+1)\phi_x^2,\ u_1, \ldots, u_k\right)$$

$$= -\frac{(2n-k+1)(2n-k-2m)(2n-k+2m+2)}{4(2n-k+2)}\phi_x^2 A_k^{n-1}\left(u_0 = \right.$$

$$-m(m+1)\phi_x^2,\ u_1, \ldots, u_k\right) + \frac{(4n-k+2)(2n-1)!!}{(2n-k+2)2^n n!}\prod_{l=0}^{n-1}(m-1)(m+1+1)\phi_x^{2n}u_k$$

$$+\overline{A}_k^{(n)}\left(u_0 = -m(m+1)\phi_x^2,\ u_1, \ldots, u_{k-1}\right) . \tag{5.6.30}$$

Because $A_k^{(0)} = u_k$, the above equation implies that

$$A_k^{(n)}\left(u_0 = -m(m+1)\phi_x^2,\ u_1, \ldots, u_k\right)$$

$$= P(n,m,k)\phi_x^{2n}u_k + \overline{A}_k^{(n)}\left(u_0 = -m(m+1)\phi_x^2,\ u_1, \ldots, u_{k-1}\right) . \tag{5.6.31}$$

The recursion relation of $P(n,m,k)$ are

$$P(n,m,k) = -Q(n,m,k)P(n-1,m,k) + S(n,m,k)\prod_{l=0}^{n-1}(m-1)(m+1+1) , \qquad (5.6.32)$$

where

$$Q(n,m,k) = \frac{(2n-1+1)(2n-k-2m)(2n-k+2m+2)}{4(2n-k+2)}$$

$$\qquad (5.6.33)$$

$$S(n,m,k) = \frac{(4n-k+2)!(2n-1)!!}{2^n(2n-k+2)n!}$$

with

$$P(0,m,k) = 1 .$$

Therefore, one can successfully determine Q's and P's and hence the coefficients $A_k^{(n)}$.

The $(2n+1)$th flow of the KdV hierarchy is given as

$$q_t = \left(L^n q\right)_x ,$$

which yields

$$\frac{1}{\phi^3}\sum_{k=0}^{\infty}\left[u_{k-1t}+(k-2)u_k\phi_t\right]\phi^k$$

$$= \frac{1}{\phi^{2n+3}}\sum_{k=0}^{\infty}\left[A_{k-1x}^{(n)}+(k-2n-2)\phi_x A_k^{(n)}\right]\phi^k ,$$

which implies that

$$(k-2n-2)\phi_x A_k^{(n)}\left(u_0,u_1,\ldots,u_k\right) = u_{k-2n-1,t}+(k-2n-2)u_{k-2n}\phi_t$$

$$-A_{k-1x}^{(n)}\left(u_0,u_1,\ldots,u_{k-1}\right) . \qquad (5.6.34)$$

It is to be noted that equation (5.6.30) can be rewritten as

$$A_k^{(n)}\left(u_0 = -m(m+1)\phi_x^2, \ u_1,\ldots,u_k\right) = P(n,m,k)\phi_x^{2n}u_k$$

$$+A_k^n\left(u_0 = -m(m+1)\phi_x^2, \ u_1,\ldots,u_{k-1}, \ u_k = 0\right) , \qquad (5.6.35)$$

which in conjunction with the above equation gives a recursion relation for u_k:

$$(k-2n-2)P(n,m,k)\phi_x^{2n+1}u_k = u_{k-2n-1,t}$$

$$+(k-2n-2)u_{k-2n}\phi_t - A_{k-1}^{(n)}\left(u_0 = -m(m+1)\phi_x^2, \ u_1,\ldots,u_{k-1}\right)$$

$$-(k-2n-2)\phi_x A_k^{(n)}\left(u_0 = -m(m+1)\phi_x^2, \ u_1,\ldots,u_{k-1}, \ u_k = 0\right) . \qquad (5.6.36)$$

This indicates that for the branch $u_0 = -m(m+1)\phi_x^2$, the resonances occur at the 0s of the polynomial in k

$$Q(n,m,k) = (k-2n-2)P(n,m,k) . \qquad (5.6.37)$$

The principal branch is

$$u_0 = -2\phi_x^2 ,$$

where the location of resonances are given at the 0s of

$$Q(n,1,k) = (k+1)[(k-5)(k-7)\ldots(k-2n-1)]$$

$$[(k-2)(k-4)\ldots(k-2n+2)](k-2n-2)(k-2n-4) , \qquad (5.6.38)$$

that is at

$$k = -1, 2, 4,\ldots , \quad (2n-2), 5, 7,\ldots , \quad (2n+1) , \quad (2n+2) , \quad (2n+4) .$$

The previous equations also give us

$$K = 0 , \quad u_0 = -2\phi_x^2 ,$$

$$K = 1 , \quad u_0 = 2\phi_{xx} \qquad\qquad\qquad (5.6.39)$$

$$K = 2 , \quad \text{compatibility}$$

$$A_2^n\left(u_0 = -2\phi_x^2, \ u_1 = 2\phi_{xx}, \ u_2\right) = 0 .$$

Now, as in the case of a single KdV equation, truncate by demanding that $u_4 = u_5 = \ldots u_{2n-1} = u_{2n+1} = u_{2n+2} = u_{2n+4} = 0$ along with $u_3 = u_{2n} = 0$. This leads to

$$u_{2t} = \left(L^n u_2\right)_x ,$$

so that the truncated expansion is a Backlund transformation. For $K = 3$, we also get

$$\left[\frac{\psi_{xx}}{\psi} + u_2 \right]_x = 0 .$$ (5.6.40)

And according to the equations, one obtains for $k = 4, \ldots, 2n - 1$

$$A_k^n\left(u_0 = -2\phi_x^2, \ u_1 = 2\phi_{xx}, \ u_2, \ u_3 = 0, \ldots, u_k = 0 \right)$$

$$= \sum_{i=0}^{k-2} g_i^{(n,k)}(\psi, U_2) \frac{d^i}{dx^i} \left[\frac{\psi_{xx}}{\psi} + u_2 \right].$$ (5.6.41)

With the help of (5.6.36), it follows that

$$\frac{d}{dx} \sum_{i=1}^{k-3} g_i^{(n,k-1)}(\psi, u_2) \frac{d^i}{dx^i} \left[\frac{\psi_{xx}}{\psi} + u_2 \right]$$

$$+ (k - 2n - 2)\phi_x \sum_{i=1}^{k-2} g_i^{(n,k)}(\psi, u_2) \frac{d^i}{dx^i} \left[\frac{\psi_{xx}}{\psi} + u_2 \right] = 0 .$$ (5.6.42)

For $k = 2n$, using the previously obtained values of u_0, u_1 etc. leads to

$$2u_0\phi_t + A_{2n-1}^n\left(u_0 = -2\phi_x^2, \ u_1 = +2\phi_{xx}, \ u_2, \ u_3 = 0, \ldots, u_{k-1} = 0 \right)$$

$$-2\phi_x A_{2n}^n\left(u_0 = -2\phi_x^2, \ u_1 = 2\phi_{xx}, \ u_2, \ u_3 = 0, \ldots, \ u_{2n} = 0 \right) = 0$$ (5.6.43)

and

$$-8\psi^4 D^{-1} \left\{ \psi \left[\psi_t - \sum_{i=0}^{n} \lambda^{n-i} \left(\frac{1}{2} B_{ix}(u_2) - B_i(u_2)\psi_x \right) \right] \right\}$$

$$+ \frac{d}{dx} \sum_{i=1}^{2n-3} g_i^{(n,2n-1)}(\psi, u_2) \frac{d^i}{dx^i} \left[\frac{\psi_{xx}}{\psi} + u_2 \right] = 0 .$$ (5.6.44)

Using the truncation conditions in (5.6.30) with $k = 2n + 1$, one obtains

$$u_{0t} - u_1\phi_t - A_{2nx}^n\left(u_0 = -2\phi_x^2, \ u_1 = 2\phi_{xx}, \ u_2, \ u_3 = 0, \ldots, u_{2n} = 0 \right)$$

$$+ \phi_x A_{2n+1}^n\left(u_0 = -2\phi_x^2, \ u_1 = 2\phi_{xx}, \ u_2, \ u_3 = 0, \ldots, u_{2n+1} = 0 \right) = 0 .$$ (5.6.45)

When equation (5.6.40) is used in this relation, we get

$$8\psi^3\left[\psi_t - \sum_{i=0}^{n}\lambda^{n-1}\left[\frac{1}{2}B_{ix}(u_2) - B_i(u_2)\psi_x\right]\right.$$
$$\left. -8\psi\psi_x D^{-1}\left\{\psi\left[\psi_t - \sum_{i=0}^{n}\lambda^{n-i}\left(\frac{1}{2}B_{ix}(u_2)\psi - B_i(u_2)\psi_x\right)\right]\right\}\right] = 0 \ . \tag{5.6.46}$$

On the other hand, for $k = 2n+2$, one gets

$$u_t - A_{2n+1x}^n\left(u_0 = -2\phi_x^2,\ u_1 = 2\phi_{xx},\ u_2,\ u_3 = 0,...,u_{2n+1} = 0\right) = 0 \ ,$$

which when using (5.6.24) becomes

$$\frac{d}{dx}\left\{\psi\left[\psi_t - \sum_{i=0}^{n}\lambda^{n-1}\left(\frac{1}{2}B_{ix}(u_2) - B_i(u_2)\psi_x\right)\right]\right\} = 0 \ . \tag{5.6.47}$$

Using equations (5.6.34) and (5.6.40) and the truncation conditions, we get

$$\phi_x A_{2n+3}^n\left(u_0 = -2\phi_x^2,\ u_1 = 2\phi_{xx},\ u_2,\ u_3 = 0,...,u_{2n+2} = 0,\ u_{2n+3}\right)$$
$$= u_{2t} - A_{2n+2}^n\left(u_0 = -2\phi_x^2,\ u_1 = 2\phi_{xx},\ u_2,\ u_3 = 0,...,u_{2n+2} = 0\right)$$
$$= u_{2t} - \left(L^n u_2\right)_x = 0 \ , \tag{5.6.48}$$

which implies

$$u_{2n+3} = 0 \ .$$

Again, for $k = 2n+4$, the same equation yields

$$A_{2n+4}^n\left(u_0 = -2\phi_x^2,\ u_1 = 2\phi_{xx},\ u_2,\ u_3 = 0,...,u_{2n+4} = 0\right) = 0 \ . \tag{5.6.49}$$

So the final consequences of all the previous conditions yield

$$A_k^n\left(u_0 = -2\phi_x^2,\ u_1 = 2\phi_{xx},\ u_2,\ u_3 = 0,...,u_{k-1} = 0,\ u_k\right) = 0 \tag{5.6.50}$$

and

$$u_k = 0 \ , \quad k > 2n+5 \ .$$

One can finally infer that if ψ is a function satisfying the following two equations

$$\psi_{xx} + (\lambda + u_2)\psi = 0$$

$$\psi_t = \sum_{i=0}^{n} \left[\frac{1}{2} B_{ix}(u_2) - B_i(u_2)\psi_x\right]\lambda^{n-1},$$

(5.6.51)

then the required consistency conditions are satisfied. The Backlund transformation has the usual form:

$$q = 2\frac{\partial^2}{\partial x^2}\ln\phi + u_2$$

$$\phi_x = \psi^2$$

(5.6.52)

$$u_{2t} = \left(L^n u_2\right)_x, \qquad q_t = \left(L^n q\right)_x.$$

The above analysis clearly shows that when implementing the Painleve analysis to a hierarchy, only the combinatorial part gets more complicated. The basic features remain the same. The treatment can also be given for the general AKNS problem. The treatment given above is taken from a paper by Newell, Tabor, and Zang. There are many other important issues for which we refer the reader to this seminal paper.

5.7 Lax type representation of Painleve equation

Long before the advent of the WTC approach of Painleve test, Flaschka[73] showed that it is possible to represent the Painleve class of ODE in the form of a Lax equation. Such a representation shows a close linkage of the Painleve class to the integrable PDEs via various similarity reductions. In this connection he also observed that there are some nonlinear ODEs that have no relation to the soliton-bearing nonlinear systems yet can have a commutator representation. This section gives the reader a glimpse of this interesting and important observation.

Consider the operator

$$L = D^2 + f, \quad D = d/dx$$

and

$$B = D^3 + (3/2f + x/2)D + 3/4f$$

and compute

$$[L, B] = L \ . \tag{5.7.1}$$

One then arrives at

$$f''' + 6ff' - 4f + 2xf' = 0 \ .$$

This nonlinear ODE is obtained from the KdV equation

$$q_t + 3qq_{x'} + 1/2q_{x'x'x'} = 0 \tag{5.7.2}$$

if one searches for a similarity solution of the form

$$q(x',t) = \left[(3t)^{-2/3} f\!\left(x'/(3t)^{1/3}\right)\right] \ . \tag{5.7.3}$$

In this connection it is important to recall that the Zakharov-Shabat equation in one space variable can be written in the Lax form,

$$[L, A] + L_t = 0 \ , \tag{5.7.4}$$

whereas the stationary solutions of this equation obey

$$[L, B] = 0 \ , \tag{5.7.5}$$

implying the existence of a polynomial relation,

$$Q(l, b) = 0$$

such that

$$Q(L, B) = 0 \ .$$

A host of researchers from the Russian School have shown how the analysis of the Baker-Akhiezer function associated with the above algebraic curve can lead to a detailed analysis of the corresponding solution.

The observation mentioned above is intriguing because equation

$$[L, B] = L \tag{5.7.6}$$

is also a consequence of the Lax pair. Consider the eigenvalue problem

$$[L - \lambda]v = 0 . \tag{5.7.7}$$

Differentiating with respect to λ one gets

$$[L - \lambda]\frac{\partial v}{\partial \lambda} = v . \tag{5.7.8}$$

But if one applies $[L, B] = L$ to an eigenvector v, one gets

$$LBv - \lambda Bv - \lambda v = 0$$

or

$$(L - \lambda)\left(\frac{1}{\lambda} Bv\right) = v . \tag{5.7.9}$$

Therefore, from equations (5.7.8) and (5.7.9), one gets

$$\frac{1}{\lambda} Bv = v_\lambda - \bar{v} ,$$

where $\bar{v}$ is an appropriate solution of $L\bar{v} = \lambda\bar{v}$ so that

$$\lambda v_\lambda = Bv + \lambda\bar{v} . \tag{5.7.10}$$

On the other hand, Bv can be re-expressed in terms of λ if one replaces x derivatives of v by multiples of λ by $Lv = \lambda v$. Therefore, one obtains

$$\lambda v_\lambda = B(\lambda)v , \tag{5.7.11}$$

where $B(\lambda)$ is some matrix that depends polynomially on λ.

Now study the whole situation in the light of Volterra operator and Zakharov dressing procedure. Let F denote an operator that acts on a function in the following way:

$$(\hat{F}\psi)(x) = \int_{-\infty}^{\infty} F(x, y)\psi(y)dy . \tag{5.7.12}$$

Also assume that F admits a triangular factorization,

$$1 + \hat{F} = \left(1 + \hat{K}_+\right)^{-1}\left(1 + \hat{K}_-\right). \tag{5.7.13}$$

$K_\pm$ are upper and lower triangular Volterra operator. In the original formulation of Zakharov $et\ al.$, they started with bare linear operators L_0, B_0

$$L_0 = 1D^n, \qquad B_0 = bD^m \tag{5.7.14}$$

and used the conditions

$$\left[L_0, B_0\right] = 0$$
$$\left[F, L_0\right] = 0 \tag{5.7.15}$$
$$\left[F, B_0 + \frac{\partial}{\partial t}\right] = 0 \,.$$

Since time dependence is not relevant for the self-similar solutions, the last condition is replaced by

$$\left[F, B_0 + T_0\right] = 0 \,. \tag{5.7.16}$$

To denote an operator whose coefficients will generally depend on x in a simple fashion, consider a very simple example:

$$L_0 = D^2, \qquad B_0 = D^3, \qquad T_0 = \frac{1}{2}xD \,.$$

Then equations (5.7.14), (5.7.15), and (5.7.16) lead to

$$F_{xx} - F_{yy} = 0$$
$$F_{xxx} + F_{yyy} + \frac{1}{2}xF_x + \frac{1}{2}yF_y + \frac{1}{2}F = 0 \,. \tag{5.7.17}$$

If we consider a solution of (5.7.17) as

$$F = F'(x+y) \,,$$

then setting $\xi = x + y$ gives

$$4 F_{\xi\xi\xi} + \xi F_\xi + F = 0$$

from equation (5.7.17). One may note that by imposing Jacobi identity one can always get

$$[F,[L_0,T_0]] = 0 , \qquad (5.7.18)$$

a necessary condition on T_0.

The procedure of dressing consists of assuming some simple perturbations L, B, T of the free operators L_0, B_0, T_0. The forms of L, B, T are introduced by the conditions

$$X(1 + K_+) - (1 + K_+)X_0 = \text{integral operator,}$$

where $X = L, B, T$ correspond to $X_0 = L_0, B_0, T_0$. Now it is well known that for

$$L_0 = D^2 \quad \text{and} \quad L = D^2 + q , \qquad (5.7.19)$$

one gets

$$q(x) = 2\frac{d}{dx}K(x,y) , \qquad (5.7.20)$$

the simplest formulae of the KdV inverse problem. But for $B_0 = D^3$, one gets

$$B = D^3 + \frac{3}{2}qD + \frac{3}{4}q' + r , \qquad (5.7.21)$$

with

$$q(x) = 2\frac{d}{dx}\xi_0(x)$$

$$\gamma(x) = \frac{3}{2}\frac{d}{dx}\left(\xi_1(x) + \xi_0^2(x)\right) , \qquad (5.7.22)$$

where

$$\xi_j(x) = \left(\frac{\partial}{\partial x} - \frac{\partial}{\partial y}\right)^j K(x,y)\Bigg|_{x=y} .$$

This time one also gets

$$T = T_0 = 1/2xD .$$

Note that the map $X_0 \to X$ is linear. In particular,

$$(B+T)(1+K_\pm)-(1+K_+)(B_0+T_0) = \text{integral operator.}$$

Now, following Zakharov and Shabat one can show that if equation (5.7.13) holds, then

$$L(1+K_\pm)-(1+K_\pm)L_0 = 0$$
$$(B+T)(1+K_\pm)-(1+K_\pm)(B_0+T_0) = 0 \ . \tag{5.7.23}$$

Now one can ensure condition (5.7.18) by choosing T_0 so that

$$[L_0,T_0] = L_0 \ , \tag{5.7.24}$$

whence from (5.7.18), (5.7.23), and (5.7.24) one gets

$$[L,B+T] = L \ , \tag{5.7.25}$$

which is the basic equation proposed by Flaschka. Below is a proof of the above assertion following Flaschka:

First multiply (5.7.23) on the left by T_0 and then on the right by T_0 and subtract the two so that

$$[T_0,L](1+K)+L[T_0,K] = (1+K)[T_0,L_0]+[T_0,K]L_0 \ . \tag{5.7.26}$$

Now rewrite (5.7.23) as

$$[T_0,K]+(T-T_0)(1+K)+B(1+K) = (1+K)B_0 \ . \tag{5.7.27}$$

Multiplying on the right and left by L_0, we get

$$L[T_0,K] = -L(T-T_0)(1+K) = LB(1+K)+L(1+K)B_0$$
$$[T_0,K]L = -(T-T_0)(1+K)L-B(1+K)L+(1+K)B_0L \ . \tag{5.7.28}$$

Using these equations in (5.7.26), we get

$$[T_0,L](1+K)-L(T-T_0)(1+K)-LB(1+K)+L(1+K)B_0$$
$$= (1+K)[T_0,L_0]-[T-T_0](1+K)L-B(1+K)L+(1+K)B_0L \ . \tag{5.7.29}$$

Now add $BL(1+K) = B(1+K)L_0$ to both sides and subtract

$$L(1+K)B_0 = (1+K)L_0 B_0 \tag{5.7.30}$$

to get

$$\{[T_0, L] + L(T_0 - T) + [B, L]\}(1+K)$$
$$= (1+K)[T_0, L_0] + (T_0 - T)(1+K)L_0 . \tag{5.7.31}$$

Also, from (5.7.23) we get

$$(T_0 - T)(1+K)L_0 = (T_0 - T)L(1+K) \tag{5.7.32}$$

so that equation (5.7.31) yields

$$\{[T_0, L] + [L, T_0 - T] + [B, L]\}(1+K)$$
$$= (1+K)[T_0, L_0] = (1+K)L_0 = -L(1+K) \tag{5.7.33}$$

operating with $(1+K)^{-1}$

$$[T_0, L] + [L, T_0 - T] + [B, L] + L = 0 , \tag{5.7.34}$$

which is nothing but

$$[L, B+T] = L . \tag{5.7.35}$$

One should remember in this connection that a basic ingredient in this approach is the equation of the Lax eigenfunction derived by differentiation with respect to λ.

5.8 Possibility of leading positive exponent

All the discussions presented in the previous chapters have asserted that it is necessary for the leading exponent of the WTC expansion to be either negative or 0 (logarithmic singularity). On the other hand, it has been observed that in many situations it is impossible to follow this dictum and formulate the test. Such a situation occurs in cases of derivative nonlinear Schrodinger equation, which is known to possess all the ingredients of a completely integrable system – such as Lax pair, infinite number of conservation laws, Hamiltonian structure, etc. Below, the example from the excellent work of Leo *et al.*, indicates the possibility of positive leading exponent and a truncation thereafter.

The derivative nonlinear Schrodinger equations (DNLS) can be written as

$$i q_t = -q_{xx} + i\left(rq^2\right)_x$$
$$i r_t = r_{xx} - i\left(9r^2\right)x \ .$$

$(5.8.1)$

As before, assume that q and r can be expanded about the singularity manifold $\phi(x,t)=0$ as

$$q = \phi^\alpha \sum_{j=0}^{\infty} u_j \phi^j$$
$$r = \phi^B \sum_{j=0}^{\infty} v_j \phi^j \ ,$$

$(5.8.2)$

where u_j, ϕ_j, v_j are analytic in the neighborhood of $\phi(x,t)=0$. A simple leading order analysis yields

(i) $\alpha = 1, \ \beta = -2$

(ii) $\alpha = -2, \ \beta = 1$

$(5.8.3)$

and one can easily demonstrate that no other matching is possible. Therefore, both exponents become negative integers or even fractions. Consider the first branch, and substitute the above in equation $(5.8.1)$ to get

$$i\left(u_{j-2,t} + j\phi_t u_{j-1}\right) = -u_{j-2,xx} - 2j\phi_x u_{j-1,x} - j\phi_{xx}u_{j-1} - j(j+1)\phi_x^2 u_j$$
$$+i\cdot \sum_{m+n<j-1} u_m u_n v_{j-1-m-n,x} + i\phi_x \sum_{m+n<j}(j-2-m-n)u_m u_n v_{j-m-n}$$
$$+2i \sum_{m+n<j-1} u_m u_{n,x} v_{j-1-m-n} + 2i\phi_x \sum_{m+n<j}(n+1)u_m u_n v_{j-m-n}$$

$(5.8.4)$

and

$$i\left(v_{j-2,t} + (j-3)\phi_t v_{j-1}\right) = v_{j-2,xx} + 2(j-3)\phi_x v_{j-1,x}$$
$$+(j-3)\phi_{xx}v_{j-1} + (j-2)(j-3)\phi_x^2 v_j + i \sum_{m+n<j-1} v_m v_n u_{j-1-m-n,k}$$

$$+i\phi_x \sum_{m+n<j}(j+1-m-n)v_m v_n u_{j-m-n} + 2i\sum u_{j-1-m-n}v_m v_{nx}$$

$$+2i\phi_x \sum_{m+n<j}(n-2)jv_{j-m-n}v_m v_n \; . \tag{5.8.5}$$

These recursion relations can be rewritten as

$$j\phi_x\left[(j-3)\phi_x u_j - iu_0^2 v_j\right] = f\left(u_j - 1, \; v_{j-1},...\right)$$
$$(j-3)\phi_x\left[iv_0^2 u_j + (j+2)\phi_x v_j\right] = g\left(u_{j-1}, v_{j-1},...\right) \; . \tag{5.8.6}$$

The determinant on the left side equated to 0 yields

$$j = -1, 0, 2, 3 \; .$$

So, for a fourth-order system, we have four resonances as well as the required one at $j = -1$.

Now the leading order analysis yields

$$u_0 v_0 = -2i\phi_x \; . \tag{5.8.7}$$

Therefore, either u_0 or v_0 is arbitrary. On the other hand, by solving the recursion for various values of j, we get

$$u_1 = \frac{iu_0}{2\phi_x^2}\left(\phi_t + i\phi_{xx} + 2i\frac{u_{0x}}{u_0}\phi_x\right) \tag{5.8.8}$$

$$v_1 = -2i\frac{u_{0x}}{u_0^2}$$

along with

$$u_0 v_1 + v_0 u_1 = \frac{\phi_t}{\phi_x} + i\frac{\phi_{xx}}{\phi_x} \; . \tag{5.8.9}$$

One can also deduce the following relations for $j = 2$ and $j = 3$:

$$i\left(u_{0t} + 2u_1\phi_t\right) = -u_{0xx} - 2u_1\phi_{xx} + 2u_2\phi_x^2$$

$$+\left(2iu_0^2 v_2 + 4iu_1 v_1 u_0 + 2iu_1^2 v_0\right)\phi_x$$

$$+i\left(v_0^2 v_{1x} + 2u_0 u_1 v_{0x} + 2u_0 u_{0x} v_1 + 2u_1 v_0 u_{0x}\right) \qquad (5.8.10)$$

$$i\left(v_{0t} + v_1\phi_t\right) = v_{0xx} + 2v_{1x}\phi_x - v_1\phi_{xx}$$

$$+i\left(v_0^2 u_{1x} + 2v_0 v_1 u_{0x} + 2u_1 v_0 v_{0x} + 2u_0 v_1 v_{0x}\right)$$

$$-i\left(u_2 v_0^2 + 2u_1 v_1 v_0\right)\phi_x - 4v_2\phi_x^2 - iu_0 v_1^2\phi_x \qquad (5.8.11)$$

$$i\left(u_{1t} + 3u_2\phi_t\right) = -u_{1xx} + u_2\phi_{xx} - 2u_{2x}\phi_x + i\left(u_0^2 v_2\right)_x$$

$$+i\left(u_1^2 v_0\right)_x + 2i\left(u_0 u_1 v_1\right)_x + 3iu_0^2 v_3\phi_x$$

$$+6iu_2\left(u_0 v_1 + u_1 v_0\right)\phi_x + 3iu_1^2 v_1\phi_x + 6iu_0 u_1 v_2\phi_x \qquad (5.8.12)$$

$$iv_{it} = v_{1xx} + i\left(v_0^2 u_2\right)_x + 4\left(v_2\phi_x\right)_x + 2i\left(v_0 u_1 v_1\right)_x + i\left(v_1^2 u_0\right)_x . \qquad (5.8.13)$$

Note that u_3 does not occur in (5.8.12), and (5.8.13) is identically satisfied. Also equations (5.8.10) and (5.8.11) are not independent, due to the equation (5.8.9), so either v_1 or u_2 is not determined.

One can therefore assume that

$$u_2 = 0, \quad u_0 = \alpha = \text{constant}, \quad u_3 = 0 .$$

It then follows that $v_1 = v_2 = v_3 = 0$, and it is possible to set $u_j = 0$ for $j > 2$. Therefore,

$$q = \alpha\phi + u_1\phi^2$$
$$\gamma = v_0/\phi^2 .$$

Note that the concept of truncation at the constant level does not work in this case. But still it is possible to truncate and define a Backlund transformation because u_1 and v_0 both satisfy the DLNS set

$$iu_{1t} = -u_{1xx} + i\left(v_0 u_1^2\right)_x . \qquad (5.8.14)$$

whereas,

$$u_1 = \frac{\alpha}{2\phi_x^2}(i\phi_t = \phi_{xx})$$

$$v_0 = -\frac{2i}{\alpha}\phi_x \, ,$$

(5.8.15)

and ϕ satisfies the equation

$$\frac{\partial}{\partial t}(\phi_t/\phi_x) + \frac{\partial}{\partial x}\{\phi;x\} - \frac{3}{2}\frac{\partial}{\partial x}(\phi_t/\phi_x)^2 = 0 \, .$$

(5.8.16)

Different form of truncation

The above example shows that when one of the leading exponents become positive, the concept of truncation of the Painleve expansion must be reformulated. This section discusses some recent ideas advocated by Pickering and Lou in connection with the truncation procedure. It is also important to note that this improved and new truncation procedure can lead to a new type of solution of the nonlinear integrable system.

To give an idea of the strategy involved, recapitulate the method of Painleve expansion in a conformal invariant manner. If we write the equation as

$$u_t = K[u]$$

(5.8.17)

and the expansion as

$$u = \phi^{-\alpha}\sum_{j=0}^{\infty}u_j\phi^j \, ,$$

$\phi(xt) = 0$, being the singular manifold. Usually a Painleve test can involve several branches depending upon different matching for the leading exponent and the corresponding coefficients. Here, α stands for the leading exponent. The leading coefficient client is u_0, and let the dominant terms be denoted as $\hat{K}[u]$. Then this initial set of information can always lead to a set of indices that is resonance position $R = \{\gamma_1,...,\gamma_n\}$ for any integrable system, determining the values of j for which arbitrary coefficients are to be introduced. So these facts can be summarized by collecting them as

$$\alpha, \quad u_0\hat{K}[u], \quad R = \{\gamma_1,...,\gamma_n\} \quad \text{and} \quad \beta \, ,$$

where β stands for the weight of $K[u]$. Here u is of weight α and $\partial/\partial x$ is of weight 1. The conformal invariant form of the above expansion can be written as

$$u = \chi^{-\alpha} \sum_{j=0}^{\infty} u_j \chi^j \tag{5.8.18}$$

with

$$\chi = \left(\phi_x/\phi - \phi_{xx}/2\phi_x\right)^{-1} .$$

One also defines two important functions S and C, to be homographic invariants

$$S = \frac{\phi_{xxx}}{\phi_x} - \frac{3}{2} \frac{\left(\phi_{xx}\right)^2}{\phi_x^2} \tag{5.8.19}$$

$$C = -\phi_t/\phi_x ,$$

which satisfy some differential relations discussed in Chapter 3. Since we have assumed the dominant terms to be weight β, after the series (5.8.18) is substituted in (5.8.17), we must have

$$K[u] - u_t = \chi^{-\beta} \sum_{j=0}^{\infty} Q_j \chi^j . \tag{5.8.20}$$

From the Riccati equations connecting χ_x, χ_t with S and C, it can be observed that any term of the form χ^p can lead to both χ^{p-1} and χ^{p+1} terms. It implies that, just as in the case when the leading coefficient is $-\alpha$ and dominant term is $(-\beta)$, that there may be a leading term with $+\alpha$ and the dominant terms at $(+\beta)$.

If we consider any two expansion branches characterized by $\{\alpha, v_0, \beta\}$ and $\{\alpha, u_0, \beta\}$, we may seek a solution as

$$u_T = \chi^{-\alpha} \sum_{j=0}^{\alpha+\overline{\alpha}} u_j \chi^j . \tag{5.8.21}$$

Corresponding to this we get

$$K[u] - u_t = \chi^{-\beta} \sum_{j=0}^{\beta+\beta} Q_j \chi^j . \tag{5.8.22}$$

Because the dominant terms depend only on spatial derivatives, the last term in the above expansion is

$$u_{\alpha+\bar{\alpha}} = \left(-\frac{1}{2}s\right)^{\bar{\alpha}} u_0 . \qquad (5.8.23)$$

If the leading exponents of both branches are the same, say α, then

$$u_T = \chi^{-\alpha} \sum_{j=0}^{2\alpha} u_j \chi^j$$

$$= u_0 \chi^{-\alpha} + ... + \left(-\frac{1}{2}s\right)^{\alpha} u_0 \chi^{\alpha} . \qquad (5.8.24)$$

u_T in our above equations always represents the truncated form of u. Consider the usual KdV case:

$$u_t = K[u] = \left(u_{xx} + 3u^2\right)_x ,$$

for which case, $\alpha = 2$, $u_0 = -2\phi_x^2$, $K[u] = K[u]$,

$$\beta = 5 , \qquad R = \{-1,4,6\} .$$

Here we can write

$$u_T = \chi^{-2} \sum_{j=0}^{4} u_j \chi^j \qquad u_0 v_4 \neq 0 .$$

Explicitly,

$$u_T = -2\chi^{-2} - 1/6(C+4S) - \frac{S^2}{2}\chi^2$$

where S and C are required to be constant. Next we can cite the example of the modified KdV equation written as

$$V_t = K[V] = \left(V_{xx} - 2V^3\right)_x ,$$

which has two branches:

$$\alpha = 1, \quad v_0 = \pm 1 : K[V] = K[V] \quad \beta = 4 \quad \text{and} \quad R = \{-1, 3, 4\} \ .$$

MKdV is a completely integrable system. The standard truncation is

$$V_T = \pm \chi^{-1} c \ , \quad C + S = 0 \ .$$

For different truncation, we can have the choice $\left(V_0, \overline{V}_0 \right) = \pm(1, 1)$ which yields

$$V_T = \pm \left(\chi^{-1} - \frac{1}{2} SX \right) .$$

Again S and C are required to be constant and satisfy $C + 4S = 0$, whereas the choice $\left(V_0, V_0 \right) = \pm(1, -1)$ yields

$$V_T = \pm \left(\chi^{-1} + \frac{1}{2} SX \right) .$$

The above discussion indicates that if a Painleve expansion is performed in some variable ξ and if ξ satisfies two Riccati equations in (x, t), then it is always possible to modify the truncation procedure and obtain a different class of solution. This idea was used by Lou to develop a more generalized truncation procedure. Suppose we have $2N + 2$ functions S_i, Y_i, and ξ satisfies two equations of the following form

$$\xi_x = \sum_{j=0}^{N} S_j \xi^j \ ; \quad \xi_t = \sum_{j=0}^{N} Y_j \xi^j \ . \tag{5.8.24}$$

One can then observe that there must be a set of $2N - 1$ constraints on S_i, Y_i:

$$S_{nt} - Y_{nx} + \sum_{j=1}^{n+1} j \left(S_j Y_{n+1-j} - Y_j S_{n+1-j} \right) = 0 \quad n = 0, 1, \ldots, N \tag{5.8.25}$$

$$\sum_{j=n+1-N}^{N} j \left(S_j Y_{n+1-j} - Y_j S_{n+1-j} \right) = 0 \quad n = N+1, \ N+2, \ldots, 2N-2. \tag{5.8.26}$$

Lou's idea was to use ξ as the variable for Painleve expansion, that is

$$u = \xi^{+\alpha} \sum_{j=0}^{\infty} u_j \xi^j \qquad (5.8.28a)$$

with α as a negative integer.

Discuss the simplest example of Burger equation. We set $N = 3$, so that

$$\xi_x = S_0 + S_1 \xi + S_2 \xi^2 + S_3 \xi^3$$
$$\xi_t = Y_0 + Y_1 \xi + Y_2 \xi^2 + Y_3 \xi^3 \qquad (5.8.28b)$$

then the constraints among the functions S_i and Y_i are

$$S_2 Y_3 - S_3 Y_2 = 0$$
$$S_{3t} - Y_{3x} + 2S_3 Y_1 - 2S_1 Y_3 = 0$$
$$S_{2t} - Y_{2x} + S_2 Y_1 - S_1 Y_2 + 3S_3 Y_0 - 3Y_3 S_0 = 0$$
$$S_{1t} - Y_{1x} + 2S_2 Y_0 - 2Y_2 S_0 = 0$$
$$S_{0t} - Y_{0x} + S_1 Y_0 - Y_1 S_0 = 0 . \qquad (5.8.29)$$

Therefore, there are five conditions on eight functions, and the arbitrariness of ξ is still there. The Burger equation is written as

$$u_t - 2uu_x + u_{xx} = 0 .$$

We now set

$$u = \xi^\alpha \sum_{j=0}^{\infty} u_j \xi^j , \qquad \alpha = -1, \ u_0 = S_0 ,$$

whence the recursion relations for u_j are given as

$$(j+1)(j-2)u_j = f_j(u_k, jk = 0, 1, 2, ..., j-1) . \qquad (5.8.30)$$

These f_j's are quite complicated functions of the variables $u_0, u, u_2, ..., u_{j-1}$. The resonances are located at $j = -1, 2$. As usual, $j = -1$ corresponds to the arbitrariness of ξ, and the resonance condition $f_2 = 0$ at $j = 2$ should be satisfied identically. The equations resulting for $j = 1$ and $j = 2$ from (5.8.30) are

$$-2u_0 u_{0x} + 2u_0^2 S_1 + 2u_1 u_0 S_0 + 3u_0 S_0 S_1$$
$$-u_0 S_{0x} - u_0 Y_0 - 2u_{0x} S_0 = 0 \tag{5.8.31}$$

and

$$-u_0 S_{1x} + u_0 S_1^2 + 2u_0 S_0 S_2 + u_{0t} + u_{0xx}$$
$$-u_0 Y_1 - 2u_{0x} S_1 + 2u_0^2 S_2 - 2u_0 u_{1x} - 2u_1 u_{0x} + 2u_1 u_0 S_1 = 0 . \tag{5.8.32}$$

Using $u_0 = -S_0$, we get

$$u_1 = \frac{1}{2S_0}(S_{0x} - S_0 S_1 + Y_0) . \tag{5.8.33}$$

Therefore, the resonance condition at $j = 2$ becomes

$$S_{0t} - Y_{0x} + Y_0 S_1 - S_0 Y_1 = 0 . \tag{5.8.34}$$

Equation (5.8.34) is the new consistency condition, or in other words, the resonance condition $f_2 = 0$ is satisfied identically. One can say then that the Painleve property of the Burger equation is re-obtained in the new expansion. In contrast to the situation of the Pickering procedure, here the derivative operator equation (5.8.28) ∂_x or ∂_t possesses different degrees in the negative and positive direction. In the negative direction, the operator has degree one while it possesses degree two in the positive direction. Therefore, the balance condition in the new truncated expansion is always different in the negative and positive direction. So from the leading order analysis, one easily finds that the expansion should be

$$u = u_0 / \xi + u_1 + u_2 \xi + u_3 \xi^2 . \tag{5.8.35}$$

Substituting this in the Burger equation and equating coefficients of ξ^α ($x = -3, -2, ..., 5, 6$), we get seven conditions. After considerable computation, they show that

$$\xi_x = K_1\left(-16+6\xi+9\xi^2+\xi^3\right)$$

$$\xi_t = K_0\left(-16+6\xi+9\xi^2+\xi^3\right) \tag{5.8.36}$$

$$u = 16\frac{K_1}{\xi} + \left(\frac{K_0}{2K}-3K_1\right)+12K_1\xi+2K_1\xi^2 \ .$$

Equation (5.8.36) can be integrated explicitly

$$(\xi-1)^2\,(\xi+8)/(\xi+2)^3 = C_1\exp(54\eta) \qquad n = K_1x+K_0t \ ,$$

so we can also find an explicit solution.

Conclusion

This concluding chapter has discussed some old and new concepts connected with either Painleve analysis or Painleve equation. Some of the procedures may be adopted to explore new horizons of Painleve test.

References

[1] Krzysztof, M. (1976) Opening address, *Mathematical Physics and Physical Mathematics*, Holland: D. Reidel Publishing Co.

[2] Ames, W.F. (1976) Some ad-hoc techniques for nonlinear partial differential equations, *Math. Phys. and Phys. Math.*, Holland: D. Reidel Publishing Co.

[3] Ames, W.F. (1972) Analytic techniques and solutions, *Nonlinear Partial Differential Equations in Eng. (Vol. II)*, New York: Academic Press.

[4] Ames, W.F. (1965) The origin of nonlinear partial differential equations, *Nonlinear Partial Differential Equations in Eng.*, New York: Academic Press.

[5] Zabusky, N.J. and Kruskal, M.D. (1965) Interactions of solitons in a collisionless plasma and the recurrence of initial states, *Phys. Rev. Lett.*, **15**, 240.

[6] Drazin, P.G. (1983) *Solitons*, Cambridge: Cambridge University Press.

[7] Fermi, E., Pasta, J., and Ulam, S. (1974) Studies of non-linear problems, 1, Los Alamos Rep. LA1940, 1955, reprod. in *Nonlinear Wave Motion* (Newell, A.C., Ed.), Providence, R.I: A.M.S.

[8] Ablowitz, M.J. (1978) Lectures on the inverse scattering transform, *Stud. in Appl. Math.*, **58**, 17.

[9] Airy, G. (1945) Tides and waves, Art. 208, in *Encycl. Metrop.*

[10] Korteweg, D.J. and de Vries, G. (1985) *Philos.*, **39**, 422.

[11] McCown, J. (1891) On the solitary wave, *Philos.*, **31**, 45.

[12] McCown, J. (1894) On the highest wave of permanent type, *Philos.*, **38**, 351.

[13] Gardner, C.S. and Morikawa, G.K. (1960) Similarity in the asymptonic behaviour of collision free hydromagnetic waves and water waves, New York Univ., *Courant Inst. Math. Sci. Res. Rep.* NYO - 9082.

[14] Lighthill, J. (1978) *Waves in Fluids*, Cambridge: Cambridge University Press.

[15] Landau, L.D. and Lifshitz, E.M. (1959) *Fluid Mechanics*, London: Pergamon.

[16] Whitham, G.B. (1974) *Linear and Nonlinear Waves*, New York: John Wiley & Sons.

[17] Miura, R.M. (1976) The Korteweg de Vries equation: a survey of results, *SIAM Rev.*, **18**, 412.

[18a] Scott, A.C., Chu, F.Y.E., and McLaughlin, D.W. (1973) The soliton: a new concept in applied science, *Proc. IEEE*, **61**, 1443.

[18b] Bullough, R.K. and Caudrey, P.J. (Ed.) (1980) *Solitons*, Springer Verlag.

[18c] Newell, A.C. (1985) Solitons in mathematics and physics, Society for Industrial and Applied Mathematics, Philadelphia.

[19] Tessos, B. (1982) On the analytical structure of chaos in dynamical systems, in *Dyn. Syst. and Chaos Proc.*, Sitges (Ed. L. Garrido), Lecture Notes in Physics, Vol. 179, Springer-Verlag.

[20] Ramani, A., Dorizzi, B., Grammaticos, B., and Hietarinta, J. (1986) *Phys.*, **18D**, 171.

[21] Helleman, R.H.G. (1981) in *Fundamental Problems in Statistical Mechanics*, Vol. 5, North Holland.

[22] Berry, M.V. (1978) in Topics in Nonlinear Dynamics, A.I.P. Conf. Proc., Vol. 46, A.I.P.

[23] Lichtenberg, A.J. and Lieberman, M.A. (1982) Regular and Stochastic Motion.

[24] Ford, J. (1975) in *Fundamental Problems in Statistical Mechanics*, Vol. 3, North Holland.

[25] Chirikov, B.V. (1979) *Phys. Rep.*, **52**, 265.

[26] Percival, I. and Richards, D. (1985) *Introduction to Dynamics*, Cambridge: Cambridge University Press.

[27] Bluman, G.W. and Cole, J.D. (1974) *Similarity Methods for Differential Equations*, New York: Springer.

[28] Ibrigimov, N.H. (1954) *Transformation Groups Applied to Mathematical Physics*, Holland: D. Reidel Publishing Co.

[29a] Gelfand, I.M. and Fomin, S.V. (1963) *Calculus of Variations,* Englewood Cliffs, N.J.: Prentice Hall.

[29b] Miura, R.M., Gardner, C.S., and Kruskal, M.D. (1968) Korteweg de Vries equation and generalizations, II, Existence of conservation laws and constants of motion, *J. Math. Phys.*, **9**, 1204.

[29c] Miura, R.M. (1968) *J. Math. Phys.*, **9**, 1202.

[30a] Magri, F. (1978) *J. Math. Phys.*, **19**, 1156.

[30b] Gardner, C.S. (1971) *J. Math. Phys.*, **12**, 1548.

[30c] Gel'fand, I.M. and Dikki, L.A. (1975) *Russ. Math. Surveys*, **30**, 77.

[31] Lax, P.D. (1968) Integrals of nonlinear equations of evaluation and solitary waves, *Commun. Pure Appl. Math.*, **21**, 674.

[32a]   Ablowitz, M.J. (1978) Lectures on the inverse scattering transform, *Stud. in Appl. Math.*, **58**, 17.

[32b]   Ablowitz, M.J. and Segur, H. (1981) *Solitons and the Inverse Scattering Transform*, Philadelphia, SIAM.

[32c]   Zakharov, V.E. and Shabat, A.B. (1972) *Sov. Phys. JETP*, **34**, 62.

[33a]   Wahlquist, H. and Estabrook, F.B. (1975) *J. Math. Phys.*, **16**, 1; (1973) *Phys. Rev. Lett.*, **37**, 1386.

[33b]   Roy Chowdhury, A. (1988) *Prolongation Structure in One and Two Dimensions*, Springer Series in Nonlinear Dynamics; in solitons introduction and application, (Ed. M. Lakshmanan), Proceedings of the Winter School, Bharathidasan University, Tiruchirapalli, India, January 5-17, 1987.

[33c]   Coronoes, J. (1979) Some Heuristic Comments on Soliton, Lecture Notes in Physics, Vol. 810, Berlin: Springer.

[33d]   Shadwick, W.F. (1980) *J. Math. Phys.*, **21**, 454.

[33e]   Dodd, R.K. and Fordy, A. (1983) *Proc. Roy. Soc.*, **A385**, 389.

[33f]   Dodd, R.K. and Gibbon, J.D. (1978) *Proc. Roy. Soc.*, **A359**, 411.

[34a]   Fokas, A.S. and Ablowitz, M.J.   The Inverse Scattering Transform for Multidimensional (2+1) Problems, Lecture Notes in Physics, 189, *Nonlinear Phenomena* (Ed. K.B. Wolf), Springer.

[34b]   Manakov, S.V., Novikov, P., Zakharov, V.E., and Pitaevsky, S. (1984) Theory of Solitons, New York: Consultants Bureau.

[34c]   Mikhailov, A. (1981) *Phys.*, **3D**, 73.

[34d]   Chudnovsky, D. and Chudnovsky, V. (Ed.) (1982) Riemann Problem, Seminar Report, *Lecture Notes in Mathematics, Vol. 925*, Springer-Verlag.

[34e]   Leo, M., Leo, R.A., Soliani, G., and Martina, I. (1986), *Inverse Problems*, **2**, 95.

[35a]   Miura, R.M. (Ed.) (1976) *Backlund Transformations, Vol. 515*, Heidelburg: Springer.

[35b]   Dold, A. and Echmann, B. (Ed.) (1976) *Lecture Notes in Mathematics, Vol. 515*, Springer Verlag.

[35c]   Zakharov, V.E. and Faddeev, L. (1972) KdV equation: a completely integrable Hamiltonian system, *Functional Anal. Appl.*, **5**, 280.

[35d]   Flaschka, H. and Newell, A.C. (1975) Integrable systems of nonlinear evolution equations, in *Dynamical Systems, Theory and Applications* (J. Moser, Ed.), Berlin: Springer.

[36a]   Ablowitz, M.J., Ramani, A., and Segur, H. (1978) Nonlinear evolution equations and ODE's of Painleve type, *Lett. Nuovo. Cimento*, **23**, 333.

[36b] Ablowitz, M.J., Ramani, A., and Segur, H.   (1980)   A connection between nonlinear evolution equations and ODE's of P-type, *J. Math. Phys.*, **321**, 715.

[37a] Weiss, J., Tabor, M., and Carnevale, G. (1983) *J. Math. Phys.*, **24**, 522.

[37b] Weiss, J. (1983) *J. Math. Phys.*, **24**, 1405.

[37c] Weiss, J. (1984) *J. Math. Phys.*, **25**, 13.

[37d] Weiss, J. (1984) *J. Math. Phys.*, **25**, 2226.

[37e] Weiss, J. (1984) *Phys. Letts.*, **102A**, 329.

[37f] Weiss, J. (1984) *J. Math. Letts.*, **105A**, 389.

[37g] Weiss, J. (1985) *J. Math. Phys.*, **326**, 258.

[37h] Weiss, J. (1985) *J. Math. Phys.*, **26**, 2174.

[37i] Weiss, J. (1986) *J. Math. Phys.*, **27**, 2647.

[37j] Weiss, J. (1986) *J. Math. Phys.*, **27**, 2647.

[37k] Weiss, J. (1983) *Circulant Matrices*, New York: John Wiley & Sons.

[37l] Weiss, J. (1987) *J. Math. Phys.*, **28**, 2025.

[37m] Weiss, J.   (1989)   *Solitons in Nonlinear Optics and Plasma Physics*, IMA, in *Mathematics and its Applications*, Springer-Verlag.

[37n] Weiss, J.   (1988)   Backlund transformation, focal surfaces and the two-dimensional Toda Lattice, *Phys. Lett. A*.

[38a] Hirota, R. *Lecture Notes in Phys. Vol. 515*, Springer-Verlag.

[38b] Steeb, W.H., Kloke, M., and Spieker, B.H.   (1984)   *J. Phy. A: Math. Gen.*, **17**, L825.

[38c] Gibbon, J.D., Radmore, P., Tabor, M., and Woodstud, D. (1985) *Appl. Math.*, **72**, 39.

[38d] Hirota, R. (1972) Exact solution of the Sine-Gordon equation for multiple collisions of solitons, *J. Phys. Soc. Japan*, **33**, 1459.

[38e] Goudrey, P.J., Eilbeck, J.C., Gibbon, J.D., and Bullough, R.K.   (1973)   Exact multisoliton solutions of the SIT and Sine-Gordon equations, *Phys. Rev. Lett.*, **30**, 237.

[38f] Hirota, R. and Suzuki, K. (1973) Theoretical and experimental studies of lattice solitons in nonlinear lumped networks, *Proc. IEEE*, **61**, 1483.

[39a] Hille, E.   (1976)   *Ordinary Differential Equations in the Complex Plane*, New York: John Wiley & Sons.

[39b] Ince, E.L. (1956) *Ordinary Differential Equations*, New York: Dover.

[39c] Davis, H.T.   (1962)   *Introduction to Nonlinear Differential and Integral Equations*, New York: Dover.

[39d] Forsyth, A.R. (1906) *Theory of Differential Equations*, New York: Dover.

[40a]   Kovalevskaya, S. (1989) *Acta Math.*, **12**, 177; (1989), **14**, 81.

[40b]   Cooke, R. (1984) *The Mathematics of Sonya Kovalevskaya*, New York: Springer.

[41]   Fokas, S. and Ablowitz, M.J. (1982) *J. Math. Phys.*

[42a]   Ramani, A., Dorizzi, B., and Grammaticos, B. (1982) *Phys. Rev. Lett.*, **49**, 1539.

[42b]   Dorizzi, B., Grammaticos, B., and Ramani, A. (1983) *J. Math. Phys.*, **24**, 2282.

[42c]   Grammaticos, B., Dorizzi, B., and Ramani, A. (1983) *J. Math. Phys.*, **24**, 2289.

[42d]   Ranada, A.F., Ramani, A., Dorizzi, B., and Grammaticos, B. (1985) *J. Math. Phys.*, **26**, 708.

[42e]   Graham, R., Rockasrts, D., and Tel, T. (1985) *Phys. Rev. A.*, **31**, 3364.

[43]   Roy Chowdhury, A., and Chanda, P.K. (1986) *J. Math. Phys.*, **27**, 707.

[44]   Roy Chowdhury, A., and Chanda, P.K. (1988) *J. Math. Phys.*, **29**, 843.

[45]   Roy Chowdhury, A., and Naskar, M. (1986) *J. Phys.*, **A19**, 3741.

[46]   Roy Chowdhury, A., and Guha, P. *Phys. Lett.*, **A134**.

[47]   Chanda, P.K., and Roy Chowdhury, A. (1988) *Int. J. Theo. Phys.*, **27**, 901.

[48]   Roy Chowdhury, A., and Chanda, P.K. (1987) *Int. J. Theo. Phys.*, **26**, 907.

[49]   Goldstein, P., and Infeld, E. (1984) *Phys. Lett.*, **103A**, 8.

[50]   Clarkson, P.A. (1985) *Phys. Lett.*, **109A**, 205.

[51]   Yoshida, H. (1983) *Celestial Mechanics*, **81**, 363.

[52]   Yoshida, H. (1983) *Celestial Mechanics*, **81**, 381.

[53]   Yoshida, H. (1989) *Phys. Lett.*, **A141**, 108.

[54]   Ziglin, S.L. (1983) *Funct. Anal. Appl.*, **16**, 181; **17**, 6.

[55]   Webb, G.M., and Zank, G.P. (1990) *Phys. Lett.*, **150A**, 14.

[56]   Lou, S. (1997) *J. Phys.*, **A30**, 4803.

[57]   Steeb, W.H., Kloke, M., Spieker, B.M. and Kunick, A. (1985) *Found. Phys.*, **15**, 637.

[58]   Conte, R. (1988) *Phys. Lett.* **A** (October).

[59]   Musette, M. (1988) Lectures at the nonlinear evolution equations' integrability and spectral methods' (como) July, 1988.

[60]   Conte, R., and Musette, M. (1988) *J. Phys.*, **A** (October).

[61]   Conte, R., Fordy, A., and Pickering, A. (1993) *Physica*, **D69**, 33.

[62]   Joshi, N., and Peterson, J.A. (1994) *Nonlinearity*, **7**, 595.

[63]   Kruskal, M.D., and Clarkson, P.A. (1992) *Stud. App. Math.*, **86**, 87.

[64]   Dorizu, B., Grammaticos, B., and Ramani, A. (1983) *J. Math. Phys.*, **24**, 2282.

[65]   Kruskal, M.P. Lectures at Painleve transeendenti, their asymptotics and physical applications - *Proceedings Ste. Adele* Quebec, Canada, September 1980. Ed. P. Winternitz and D. Levi (Plenum Press, New York 1990).

[66] Gibbon, J.D., and Tabor, M. (1985) *J. Math. Phys.*, **26**, 1956.

[67] Fokas, A., and Yortos, Y.C. (1987) *Lett. Nuovo Cinento*, **30**, 539.

[68] Gurani, V.P., and Matveev, V.I. (1985) *Sov. Math. Dokl*, **31**, 52.

[69] Ercolani, N., and Siggia, E.D. (1984) *Physica*, **D34**, 303.

[70] Roy Chowdhury, A., and Naskar, M. (1980) *J. Phys.*, **A19**, 3741.

[71] Newell, A.C., Tabor, M., and Bzeng, Y. (1990) Arizona preprint.

[72] Cariello, F., and Tabor, M. (1989) *Physica*, **39D**, 77.

[73] Flaschka, H. (1980) *J. Math. Phys.*, **21**, 1018.

[74] Ablowitz, M.J., Ramani, A., and Segur, H. (1980) *J. Math. Phys.*, **21**, 775; **21**, 1006.

[75] Lou, S. (1998) *Zeit. Fur Naturforcshung*, **53**, 251.

[76] Pickering, A. (1993) *J. Phys.*, **A26**, 4395.

[77] Steeb, W.H., Grauel, A., Kolke, M., and Spieker, B.M. (1985) *Physica Scripta*, **31**, 5.

[78] Grammaticos, B., and Ramani, A. Integrability and how to defect it (Preprint).

[79] Grammaticos, B., Ramani, A., and Papageorgiou, V. (1991) *Phys. Rev. Lett.*, **67**, 1825.

[80] Roy Chowdhury, A., and Guha, P. (1988) *Phys. Lett.*, **A134**, 115.

[81] Jimbo, M., and Miwa, T. (1981) *Physica*, **D2**, 407; **D4**, 26.

[82] Papageorgiou, V.G., Nijhoff, F.W., Grammaticos, B., and Ramani, A. (1992) *Phys. Lett.*, **A164**, 57.

[83] Musette, M., and Conte, R. (1991) *J. Math. Phys.*, **32**, 1450.

[84] Lou, S. (1997) *Phys. Rev. Lett.*, December.

[85] Roy Chowdhury, A., and Banerjee, R.S. (1989) *Jur. Phys. Soc.* (Jpn), **58**, 407.

[86] Mathieu, P. (1988) *Phys. Lett.*, **A128**, 169.

[87] Roy Chowdhury, A., and Naskar, M. (1987) *J. Math. Phys.*, **28**, 1809.

[88] Van Moerbeke, P. (1989) Geometry of the Painleve Analysis Lecture at the Finite Dimensional integrable nonlinear dynamical system held at Johannesburg, South Africa (World Scientific, 1989).

[89] Adler, M., and Moerbeke, P. (1982) *Invent. Math.*, **67**, 297.

[90] Adler, M., and Moerbeke, P. (1984) *Proc. Nat. Acad. Sci.* (USA), **81**, 4613.

[91] Knecht, M., Pasquier, R., and Pasquier, J.Y. *LPTHE-ORSAY* 95-96 Preprint.

[92] Fordy, A.P., and Pickering, A. (1991) *Phys. Lett.*, **A160**, 347.

[93] Fordy, A.P., and Pickering, A. (1992) in *Chaotic Dynamics: Theory and Practice*, p101-114, T. Bountis (Ed.). Plenum, New York.

[94]   Papageorgiou, V.G., Nijhoff, F.W., and Capel, H.W. (1990) *Proc. Int. Workshop on Quantum Groups*, Leningrad 1990, L.O. Faddeev and P.P. Kulish (Eds.). Springer, Berlin.

[95]   Quispel, G.R.W., Roberts, J.A.G., and Thompson, C.J. (1988) *Phys. Lett.*, **A126**, 419; (1989) *Physica*, **D34**, 183.

[96]   Roy Chowdhury, A., and Banerjee, R.S. (1989) *J. Phys. Soc.* (Jpn), **58**, 407.

[97]   Basak, S., and Roy Chowdhury, A. *Int. J. Theo. Phys.*

[98]   Roy Chowdhury, A., and Chanda, P.K. (1986) *ICTP* (Trieste), Preprint.

[99]   Roy Chowdhury, A., and Paul, S. (1983) *ICTP* (Trieste), Preprint.

[100]   Levine, G., and Tabor, M. (1988) *Physica*, **D33**, 189.

[101]   Roy Chowdhury, A., and Roy, S. *J. Phys.*, **A**.

[102]   Jimbo, M., Kruskal, M.D., and Miwa, T. (1982) *Phys. Lett.*, **92A**, 59.

[103]   Flaschka, H., and Newell, A.C. (1982) in *Nonlinear Problems Present and Future*, A.R. Bishop, D.K. Campleell, and B. Nicotaenko (Eds.). North. Holland Publishers.

[104]   Ablouitz, M.J., and Clarkson, P.A. *Solitons, Nonlinear Evolution Equations and Inverse Scattering*. London Math. Soc. V. 149, Cambridge University Press.

Index